Making Time

Studies of the Weatherhead East Asian Institute, Columbia University

The studies of the Weatherhead East Asian Institute of Columbia University were inaugurated in 1962 to bring to a wider public the results of significant research on modern and contemporary East Asia.

Making Time

Astronomical Time Measurement in Tokugawa Japan

YULIA FRUMER

The University of Chicago Press
Chicago and London

The University of Chicago Press, Chicago 60637
The University of Chicago Press, Ltd., London

Published 2018
Printed in the United States of America

27 26 25 24 23 22 21 20 19 18 1 2 3 4 5

ISBN-13: 978-0-226-51644-8 (cloth)
ISBN-13: 978-0-226-52471-9 (e-book)
DOI: 10.7208/chicago/9780226524719.001.0001

Library of Congress Cataloging-in-Publication Data

Names: Frumer, Yulia, author.
Title: Making time : astronomical time measurement in Tokugawa Japan / Yulia Frumer.
Other titles: Studies of the Weatherhead East Asian Institute, Columbia University.
Description: Chicago ; London : The University of Chicago Press, 2018. | Series: Studies of the Weatherhead East Asian Institute, Columbia University
Identifiers: LCCN 2017015017 | ISBN 9780226516448 (cloth : alk. paper) | ISBN 9780226524719 (e-book)
Subjects: LCSH: Time measurements—Japan—History. | Japan—History—Tokugawa period, 1600–1868.
Classification: LCC QB213 .F795 2018 | DDC 529/.709520903—dc23
LC record available at https://lccn.loc.gov/2017015017

♾ This paper meets the requirements of ANSI/NISO Z39.48-1992 (Permanence of Paper).

CONTENTS

NOTE ON NAMES AND TRANSLATIONS

Japanese names are customarily written last name first. Conventionally, when only one name is used, it is typically the last name. Nevertheless, this pattern is often broken in historiography of the Tokugawa period due to the abundance of professional or political lineages, all members of which adopted the same last name. To distinguish between members of the same lineage, scholars usually refer to Tokugawa-period figures by one of their pseudonyms. This book follows historiographic convention and refers to people by the names by which they are most commonly known. When there is no established convention, this text follows the customary abbreviation of full names to last names alone.

With a few exceptions (where customary translations are available), most of the title translations in this book are the author's own. As literal translations often result in convoluted or incomprehensible English, the author prioritized substance over literal precision in translating the characters. The names and titles in original characters are provided in the bibliography.

Introduction

Hunting for exotic curiosities at local markets, Western visitors to Meiji Japan (1868–1912) encountered some particularly strange objects. Obviously mechanical, they were marked with Japanese characters, and though one could guess that they were measuring devices, it was not always clear what they were supposed to measure. Some of them resembled Western clocks, but others looked more like devices designed to measure length or height. Westerners bought them to serve as elegant objets d'art that exhibited "oriental" qualities—lacquered wood, gilded panels, and exotic writing. By the beginning of the twentieth century, British collectors began to place these devices in public exhibitions. But to the collectors' amazement, when they invited Japanese embassy officials to share their knowledge of these artifacts, the diplomats claimed that they had never seen such objects before and were surprised to discover that they had come from Japan.[1]

The objects in question were antique Japanese clocks. Born of the encounter between European timekeeping technology and Japanese temporal practices, these clocks were highly valued throughout the Tokugawa (or Edo) period (1600–1868).[2] Rendered obsolete by the calendrical reform of 1873, which marked the adoption of Western methods of timekeeping, they were quickly forgotten by all but a few connoisseurs of Japan's technological history. According to one European collector, by the beginning of the twentieth century Japan had become "largely denuded of these interesting objects; and even the cultivated Japanese appeared to be ignorant of the existence of these relics of pre-revolution days."[3]

Mechanical timekeeping technology was first brought to Japan from Europe in the middle of the sixteenth century. Yet despite the general enthusiasm mechanical clocks evoked, they were deemed useless because they failed to measure time in units that fit the local timekeeping system.

According to this system, the day was divided into twelve "hours," six for daytime and six for nighttime. Because the durations of light and darkness change throughout the year, the length of these hours changed as well. So, at the height of summer, daytime "hours" lasted about two and a half modern hours each, while in the winter the situation was roughly reversed. It was only around the equinoxes that all twelve hours were more or less the same length. With seasonal changes in mind, sixteenth-century Japanese did not find the equal hours measured by mechanical clocks to be a useful to measure time.[4] Yet despite their rejection of European timekeeping conventions, they did find uses for Western mechanical timekeeping technology. In order to make mechanical timekeeping technology usable, Japanese artisans thoroughly modified the appearance and mechanism of European clocks. Adapting foreign technology to serve local conventions, they came up with designs that significantly differed from their European prototypes, and sometimes did not resemble European timepieces at all.

This reaction to the arrival of new timekeeping technology and the practices associated with it could not be more different from what happened some three hundred years later. In 1873 the Japanese government decisively discarded the country's existing temporal system in favor of the Western one, in which the day is divided into twenty-four equal hours, each divided into sixty minutes that are in turn divided into sixty seconds. This change made all earlier Japanese clocks obsolete, and forced their owners to switch to European clocks and watches.

This book explores the profound shift in attitude toward foreign technology that occurred between the sixteenth century, when European devices arrived in Japan, and 1873, when Japan finally abandoned its traditional temporal system. Before the nineteenth century, I argue, time measurement was understood to be a human-made and agreed-upon praxis—a convention. In this framework, several different methods of measuring time coexisted and could be employed for different purposes. However, engagement with Western astronomical calculation and observation methods over the course of the eighteenth century brought about a new way of conceptualizing time measurement. The new notion dictated that the time measured by astronomers was a mathematical expression of celestial motion, and therefore it was not merely *a* time, but *the* Time (with a capital T). The astronomical timekeeping system thus came to be seen as the sine qua non of Time measurement, a model for all timekeeping practices to follow. And since, by the middle of the nineteenth century, Western temporal conventions and timepieces had become associated with astronomical time measurement, they were now seen as desirable—even inevitable.

By documenting shifts in the ideas, practices, and technologies connected to time measurement in Japan over roughly three hundred years, this book offers a fresh interpretation of the ways societies evaluate and attach meanings to technologies. It shows how one society's conceptualization of timepieces evolved as a result of transformations in the broader range of associations related to the measurement of time. Once a variety of calculation methods, artifacts, and modes of visual representation transformed the associations with time measurement, the understanding of what it meant to measure time changed as well. Foreign technologies and foreign methods were thus accepted or rejected not because they more or less adequately addressed practical needs, but on the basis of how well they fit a prevailing set of norms and assumptions. In order to adapt foreign timekeeping technology to their needs, Japanese users first needed to integrate it into their web of associations related to practices of time measurement—either by modifying the technology to better suit those associations or by changing the associations to fit the technology.

Technology and Society

This claim—that the assessment of technology is tied to the broader associations it evokes—complicates the standard argument that the abandonment of the system of variable hours was a response to the Meiji period's rapid industrialization and modernization. Echoing E. P. Thompson's thesis (advanced in the 1960s) that factory work facilitated the emergence of modern, abstract, and urban time, scholars of Japanese history, too, have argued that so-called modern technologies, such as factories, railroads, and telegraphs, forged new social structures that required new timekeeping practices.[5] There are several reasons why many find this argument compelling. First, the widespread adoption of technologies such as trains and telegraphs in the Meiji period happened concurrently with certain changes in temporal practices.[6] Second, Meiji-period intellectuals explicitly propagated the idea that the Meiji period marked a clear break with the past.[7] And, finally, for a modern-day observer, it is difficult to imagine how an alien system of time measurement could possibly have functioned in the modern world, characterized as it is by down-to-the-second scheduling.

Yet none of these three reasons withstand the scrutiny of historical research. First, the concurrent adoption of modern technologies with Western timekeeping conventions during the Meiji period points only to correlation, not causation. Second, the statements of Meiji-period ideologues need to be carefully weighted against their particular aspirations and the temporal practices of their time, as I discuss in chapter 8. And third, the scheduling demands of an industrialized society rely on synchronization rather than any particular kind of temporal system.[8] Theoretically, if timepieces measuring variable hours could have been more exactly synchronized, they would have equally well served the railroad. It is true that the standards of synchronization have changed. Yet today's synchronization is enabled by technology that did not exist until quite late in the twentieth century—neither in Japan nor anywhere else.[9] In other words, our current standards of synchronicity could not have possibly motivated Meiji reformers to adopt the equal hour system, with or without trains.

If we follow David Edgerton and look at the actual use of timekeeping technology we discover that the social and technological conditions at the time of the calendrical reform of 1873—just six years after the Meiji period began—were not much different from those of the final years of the Tokugawa period.[10] When the reform was carried out—let alone when it was conceived during early nineteenth century—very little in the way of industrial transformation had taken place. Modern technologies such as steam engines were familiar to only a small minority of Japanese—the same ones who had encountered these technologies in the 1850s and '60s. At the same time the system of variable hours perfectly satisfied the scheduling and the synchronization needs of the first six years of the Meiji period as it had the final years of the Tokugawa period. Bell towers allowed everybody to know the agreed-upon time. Portable mechanical clocks that measured variable hours allowed people to follow the hours between bell strikes. Activities began and ended "by the bell." And schedules were adjusted to reflect seasonal changes in the length of hours.[11]

Thus, it is a stretch to claim that temporal reform happened *as a result* of the broad changes associated with the Meiji period.[12] Rather, the idea that Western time was desirable emerged before the management of time proved to have practical advantages in an industrialized Japan. Consequently, if we want to find an explanation for the reasoning that led to this reform we should look at the era that preceded it, the Tokugawa period.

Mechanical Clocks in Tokugawa Japan

Tokugawa Japan offers a particularly interesting case for exploring the history of technological and conceptual changes related to the measurement of time. The system of variable hours, so exotic to most modern-day readers, is intriguing in itself. To understand how people lived their lives according to hours that changed their lengths is to grapple with social and conceptual aspects of timekeeping that we often take for granted. At the same time, examining the testimonies of Tokugawa-period Japanese who found the *European* system to be bizarre and nonsensical helps us to reflect on our own assumptions and the tacit presuppositions that sustain our own system.[13]

Tokugawa Japan is not the only society that employed a system of variable hours, but it is unique in its use of such a system during a period of rapid modernization. Similar variable hours systems have also existed in China and in Europe, although by the Song period (960–1279) in China, and by the fourteenth century in Europe, variable hours had gradually given way to the equal hour systems in use today.[14] In Tokugawa Japan, however, the variable hours system was in use until the second half of the nineteenth century. It was used not only in the countryside but also in highly urban environments, in a society that was gradually developing industrial modes of production and a proto-capitalist economy.[15]

Even more noteworthy is the fact that Japanese clock-makers adapted European mechanical technology to keep hours of variable length. When Europeans first brought mechanical clocks to Japan, the Japanese did not rush to adopt the foreign timekeeping system these clocks served, but instead modified the clocks themselves. Furthermore, Japanese clock-makers found not just one, but numerous ways to adjust European technology to a variable hours system, devising a variety of designs, many of which looked nothing like their European antecedents.

The growing appreciation of the Western timekeeping system during the Tokugawa period, on the other hand, is a rare example of *voluntary* adoption of Western practices by a non-Western society. Studies on the transmission of technology from Western Europe have typically focused on scenarios in which Europeans exercised political and cultural dominance.[16] In such scenarios, European technology was imposed by occupiers, or promoted by local elites who adopted Western cultural models, as happened in Meiji Japan. In such cases, the agents of transmission assumed the superiority of Western technology. Unsurprisingly, studies that examine the transmission of technology in such situations work to dismantle the notion that there was anything inherently better in imported technology and instead tend to emphasize various modes of resistance to foreign technology or its subversive adaptation. Yet the situation in Tokugawa Japan was quite different—there was no need to *resist* European political dominancy in order to use European technology and practices in ways that locals found appropriate.

Although Japanese rulers initially welcomed the Europeans upon their arrival in the mid-sixteenth century, they soon came to see the foreigners as destabilizing and dangerous. Europeans brought not only clocks, tobacco, and exotic foods but also firearms and Christianity. They sold guns to the highest bidder, aiding forces opposed to the emerging central power. Rebellions of recent converts to Christianity troubled Japanese rulers, who, having dealt with rebellions led by militant Buddhist monks, were deeply wary of religious zeal.[17] Moreover, Japanese who traveled abroad brought back reports of European colonies in different parts of the world, suggesting that the Europeans might harbor similar plans for Japan. Consequently, the European presence in Japan was gradually circumscribed, Christianity was banned, and any expression of affinity with Europeans became suspect. By the mid-seventeenth century only the Dutch, whom the Japanese mistakenly perceived as non-imperialist and non-Christian, were allowed to remain, and even they were restricted to a small trading post on the artificial island of Dejima, built especially for this purpose off the coast of Nagasaki.[18]

If there was a power imbalance in this relationship during the Tokugawa period, it was tilted toward Japan. The Japanese government regulated trade with the Dutch East India Company (commonly known as the VOC—*Verenigde Oost-Indie Compagnie*), and dictated what and how much could be brought to Japan from Europe.[19] Censors inspected European goods for Christian propaganda, rejecting any book that contained even the slightest reference to Christianity, and local officials searched Dutch ships for contraband. During the first part of the Tokugawa period there was also a widespread sense of Japanese cultural superiority. The term used to describe the Portuguese literally meant "southern-barbarians" (*nanban*), while the term used to describe the Dutch was "red-furs" (*kōmō*).[20] As attitudes toward Europe began to change, these terms gradually lost their derogatory meanings. Yet even by the beginning of the nineteenth century, the foreigners were in no position to dictate to the Japanese the terms of cultural and economic exchange.

For Tokugawa Japanese, then, decisions to adopt European technologies were independent of political or economic pressures or the thrall of a general cultural admiration. Tokugawa scholars did not automatically accept the teachings of Western sciences: they weighed the pros and cons. Some scholars saw Western sciences as failing to reflect the ever-changing nature of the universe, while others insisted that Western methods offered some useful advantages, and yet others used Western armillary spheres all while claiming that Western cosmology was impractical.[21]

Tokugawa Japan thus affords an opportunity to explore a dynamic of scientific and technological transfer that occurred largely in the absence of political or cultural coercion. If we set aside the assumption that Western sciences and timekeeping technologies were somehow inherently better, we can investigate what it was that made these technologies and sciences *seem* "better" to Japanese scholars.[22] Thus, critical attention to the historical context of Tokugawa Japan can illuminate which particular historical and cultural factors caused the Japanese to gradually adopt European methods of dealing with time, as well as how Japanese assessments of Western science and culture had shifted in a positive direction by the middle of the nineteenth century.

Finally, Tokugawa Japan provides us with a rare opportunity to examine how foreign technology was interpreted *in the absence of knowledge of its original cultural context*. After the expulsion of Europeans from the country, most Japanese had few opportunities to experience European culture or talk to actual Europeans. One could take a long trip to Nagasaki or wait for the Europeans' annual visit to Edo, but even then there was no guarantee that an encounter would be productive.

First, there was the language issue. Japanese interpreters who worked with the Dutch had to study the language by immersion, with no guides other than their elders. Tasked with facilitating commerce, interpreters focused on verbal communication related to business and everyday activities, and the only written texts they dealt with on a daily basis were inventories. Before the end of the eighteenth century, the Japanese had no multilingual dictionaries or grammar books, and interpreters could not rely on European guidance because no Westerners stayed long enough at Dejima to acquire proper Japanese skills. The first attempt to produce a Dutch-Japanese dictionary ended in 1768, when the interpreter-in-chief, Nishi Zenzaburō, died in the middle of compiling words beginning with the letter B.[23] Only in the early nineteenth century did interpreters gain sufficient proficiency to be able to translate complex technological and scientific concepts.[24]

Moreover, though the residents of Dejima were Europeans, they weren't necessarily able to provide answers to the questions Japanese scholars attempted to ask. Periodically, European naturalists sought postings in Dejima to explore the flora and fauna of the distant archipelago. Aside from the sporadic appearance of such visitors—who included Engelbert Kaempfer (1690–1692), Karl Peter Thunberg (1775–1776), and Philip Franz von Siebold (1823–1829)—Dejima was principally occupied by merchants and sailors, many with limited education. Thus, when asked about the exotic animals portrayed in a Dutch book brought to Japan, for example, the merchants replied that they could not help because the books "included so many Latin words."[25]

Consequently, it fell to Japanese scholars and artisans to interpret and make sense of European objects, images, and texts. The European artifacts that Tokugawa Japanese encountered were detached from their original cultural context and background. They came without manuals explaining their uses and meanings, and the Westerners then in the area proved to be of little help. Whatever conclusions Japanese scholars and artisans reached about European devices, whatever potential they recognized in the structures of such objects, and whatever modifications to them they made were all based on their own rationale.

Technological and Conceptual Changes

In exploring Japanese users' and makers' interpretations of European timekeeping technologies, this book also shows how those interpretations varied and how they changed over the course of the Tokugawa period. What sixteenth- and seventeenth-century Japanese users saw in Western clocks and the Western timekeeping system contradicted their sense of the proper measurement of time. Western clocks failed to reflect seasonal changes in the amount of daylight, did not indicate dusk or dawn, and divided time into bizarre units that could not be incorporated by the local system of temporal calculation. For Japanese users, the only way to salvage mechanical timekeeping technology was to modify it greatly to make it fit inherited practices of keeping time.

In the eighteenth century, however, there was a broader shift in the conceptualization of what time measurement meant. The change began among a narrow circle of astronomers.[26] As purely algebraic computation of the calendar gave way to trigonometry, Japanese astronomers began timing segments of the paths traversed by celestial bodies. They also began drawing diagrams in which measured time stood for arcs described by moving planets, and could be represented as the angle between the two lines connecting a terrestrial observer with a celestial body at two different points. Having trained what historian of engineering Eugene Ferguson calls their "minds' eye" to see time in this new way, they began seeing time as a reflection of celestial motion.[27] As astronomical methods came to be employed in land surveying, the new conceptualization of time measurement manifested in cartography. Contemplating the use of land-surveying methods at sea, Japanese scholars suggested using Western chronometers instead of pendulum clocks. Chronometers were desirable because they were designed for stability amidst the rolling of the sea and measured time in the same units as British nautical almanacs, which offered observational data for a variety of locations all over the globe. The use of nautical almanacs drew attention to the previously neglected observation that the length of an astronomical day (i.e., the time from one noon to the next) changed slightly throughout the year. The units of Western clocks reflected the average, the so-called *time equation*, and thus, in the eyes of Japanese scholars they no longer seemed random and bizarre but rather a reflection of the physical reality of the universe.

By the mid-nineteenth century, Western clocks and the Western timekeeping system had become associated with highly esteemed sciences such as astronomy, geography, and navigation, as well as the benefits they provided in the form of calendars, maps, and other aids to commerce and defense. With such associations in mind, Western timekeeping conventions were deemed valuable, even if they still seemed outré. The effort required to study and adopt them now seemed worthwhile. At the beginning of the Meiji period, when science, commerce, and international relationships became associated with the ideology of modernization and enlightenment, Western clocks came to be seen as the material embodiment of these new ideals.

This book is part of a broader reconsideration of the changing meanings of terms and concepts in Tokugawa Japan. In recent years several scholars have rebelled against the idea—often propagated in historical dictionaries and classical-language textbooks—that changes in the meanings of many Japanese words occurred only with Meiji-period Westernization. Maki Fukuoka has shown how the modern sense of the word *shashin*, which is now translated as "photograph," evolved out of changes in the Tokugawa-period use of the term, which was applied to various modes of pictorial representation.[28] Federico Marcon has demonstrated how the concept of nature shifted during the Tokugawa period, developing into something similar to the modern concept behind the word *shizen*, even before this word was used to signify "nature" in its contemporary sense.[29]

What this book offers, beyond a description of changing conceptualizations of time measurement, is an explanation of the mechanisms that drove these changes. In so doing, it looks closely at the practices of measuring time. Time is a broad and abstract concept. Debates about the nature of time have occupied generations of philosophers, poets, religious figures, and—during the past few centuries—scientists.[30] Its measurement, on the other hand, has often been regarded as straightforward: with the commonsensical understanding that more precision is better. Such an approach presupposes that time has a certain stability and uniformity independent of human perception. This assumption, however, is itself rooted in a particular understanding of time that developed in the West, and later in Japan as well.[31] But if we take seriously the particularities of different peoples' ways of measuring time, we quickly realize that our modern adherence to a single system of timekeeping for all purposes obscures the fact that time measurement is never value-neutral.

Perceptions of time are shaped by the ways that people employ the time they measure. Thus, for example, measuring time for the purpose of astronomical calculations is different from measuring work hours. The goals of time measurement, as well as the very methods employed to achieve those goals, shape the ways people perceive the measurement of time. Looking at Tokugawa astronomers, we discover that their goals stayed fairly consistent—they timed celestial events for the purpose of improving the calendrical algorithm. However, their methods differed. As astronomers' mathematical methods changed from algebraic to trigonometric, their calculation practices changed as well (from scribbling numbers to drawing diagrams), the timepieces they used evolved, and so did the units in which they measured time. As a result astronomers began conceptualizing time differently. They no longer saw measured time as a sequence of points, but rather as discrete segments of movement that could be described in arcs and angles. Time no longer reflected cycles of recurrence of specific celestial constellations, but rather the relationship between the observer and celestial motion.

The measurement of time, in other words, was entangled in an array of social, material, and intellectual factors.[32] Astronomers measured time at the observatory in Edo's Asakusa district, surrounded by a library with books written in classical Chinese or in European languages they could not read. Geographers measured time in the course of astronomical surveying to the north of the main Japanese island, noticing how cold weather affected their instruments. Bell keepers measured time with as many as three different timepieces. Travelers, meanwhile, measured time in areas where time-bells could not be heard. For these various actors, time was not abstract, and its measurement was not just quantitative. Rather, the meanings of the acts of measuring time were rooted in particular cultural practices.

And how does all this relate to what people were *thinking* about time measurement? How did the conceptual world of a given individual relate to the surrounding social and material environment? Building on the long tradition of historians of science who have grappled with similar questions, this book examines the social and the material environment as a resource, a set of building blocks used for developing a theoretical understanding of the world.[33] In particular, I consider instruments, images, habits, etc., as a source of innumerable associations that shape one's conceptualization of the phenomena he or she encounters.[34] A person's particular associations with the measurement of time allow him or her to make judgments about, attach values to, and formulate theoretical ideas about time. As practices, instruments, and imagery varied among different people and changed over the years, personal associations have changed as well, and with them the ways people conceptualized time and its measurement. Contrary to narratives of modern abstract time, this book shows that, as used by particular people, *every* conception of time is task-oriented. While in use, the concept of time—or any concept for that matter—is never "abstract," but rather rooted in a series of "concretes."

Chapters and Themes

In order to understand changes in the perception of time measurement, this book begins by looking at the particulars of the variable hours system. Chapter 1 chronicles the development of timekeeping practices across the Tokugawa period. Challenging the common characterization of variable hours as "natural," I argue that their use required *more* regulation than an equal hours temporal system, not less.[35] As a consequence, Tokugawa society had already instituted "modern" time-related practices such as, for example, synchronized public timekeeping and the assignation of the whole country to one time zone.

Chapter 1 also reveals shortcomings of social construction theory, which fails to explain why certain practices persist even when the social conditions that purportedly support them change.[36] If practices and technologies develop in response to specific social needs—the logic of social constructivism goes—then every practice should be explainable in relation to a corresponding need. By this logic, if a society uses variable hours, it must be because there is a particular need for that temporal system—such as the demands of an agrarian life-style. The persistence of the variable hours system, then, cannot be explained other than by an appeal to the stability of the social conditions that supposedly required it. Yet the two hundred fifty years preceding the Meiji period saw great societal changes.[37] The seventeenth century witnessed rapid urbanization, with Edo growing from about twenty thousand to more than a million people in less than a century. Developing mining industries fueled both local economies and international trade, positioning Japan within a global economic network despite its ban on international travel. And the proto-capitalist and proto-industrial nature of mid-nineteenth-century business enterprises created work patterns quite similar to those of the early Meiji period. The variable hours system survived all these changes because it was maintained out of *convention.*[38] Changing social needs did affect modes of timekeeping and the use of timepieces, but the changes were introduced *within* the confines of the variable hour system, without requiring its overhaul.

The shortcomings of a strict social constructivist approach also feature in chapter 2, which focuses on Japanese clock-makers' modifications to European devices. Whereas chapter 1 calls attention to the fact that certain changes in social needs *do not* result in technological transformation, chapter 2 shows that other social needs result in transformations that take more than one shape. Japanese clock-makers found more than one way to adapt European clockwork to measure variable hours and came up with a variety of designs, all of which responded to the same need in different ways. Thus, although the need to measure hours in variable units might explain the *why*—the motivation to adjust Western clocks—it cannot explain the *how*—the process by which Japanese clock-makers arrived at a range of technological solutions. Chapter 2 claims that to understand the *how* requires us to acknowledge that—contra Latour—instruments do have history, and not only sociology.[39] Although timekeeping practices and technologies *are* socially constructed, they are driven not ony by the particular needs of a given moment, but also by longstanding conventions, habits, and imagery.

In order to understand the reasoning of Japanese clock-makers, this chapter turns to an invaluable historical source—the clocks themselves. Artifacts, as Lorraine Daston has put it, crystallize human experience.[40] As humans invent new technologies and modify existing ones, their reasoning becomes embedded in the very materiality of the objects they create.[41] Reading these devices as historical sources, particularly in conjunction with written accounts, reveals some of the motivations and thoughts of their creators.[42] Japanese clocks constitute a particularly rich source since they only barely, if at all, resemble European clocks. They feature movable digits, graphs, curvy lines, and index hands of adjustable length. These unfamiliar design elements provide us with a glimpse at what Michael Polanyi has called the "tacit dimension."[43] According to Polanyi, our actions are supported and enabled by a vast body of tacit knowledge that remains largely unnoticed and unspoken of. It is only in times of misunderstandings or miscommunications, or when something unexpected and unfamiliar occurs, that tacit assumptions become visible.[44] Japanese clocks offer us a glimpse of the unexpected and unfamiliar, and force us to ask "Why?"

Since Japanese clock-makers were not familiar with European assumptions about time and its measurement, they had to rely on their own interpretations of the material characteristics—the *thingy-ness*, as Davis Baird has called it—of the devices themselves.[45] They interpreted these material characteristics on the basis of their own existing associations with time measurement, including practices with other, nonmechanical timekeeping devices, familiar objects in their households, and customary visual representations of changes in time. Thus, they did not just ask how they could adjust the clocks to *measure variable hours*. Instead, they asked how they could use the mechanical structure to make the clock measure time *in a way that they knew*. So they designed clocks that looked like other measurement devices they had, such as rulers. They made the digits movable, just like the digits in an incense clock. And they made the clocks look like graphs of seasonal changes in the amount of daylight—a format people were used to reading. In sum, they modified the clocks according to their existing associations with the variety of ways time was measured, read, and depicted.

Chapter 3 explores the importance of changing associations to the process of knowledge transfer. Focused on the timekeeping practices of Tokugawa astronomers, this chapter shows how evolving mathematical methods produced new sets of associations relevant to the measurement of time. Equipped with new associations, every generation of Tokugawa astronomers came up with new ways of thinking about what it means to measure time. In particular, they began seeing time in terms of celestial arcs and angles, rather than as sequences of temporal points. In order to match their practice with their evolving concepts of time measurement, astronomers adopted new ways of measuring time and designed new kinds of timepieces. Gradually, as their practice came to include more and more European elements—Western trigonometry, calculation using diagrams, terrestrial and celestial maps, sextants, and pendulums—their associations changed as well. Employing methods similar to those of their European colleagues, Tokugawa astronomers formed new assumptions about the nature of the universe that were similar to assumptions characteristic of European astronomy.

Tokugawa astronomers' engagement with European astronomical practices thus highlights a thorny issue in the history of Japanese science and technology: the adoption of knowledge versus its creation. Looking at Tokugawa astronomers' practices as situated in a web of associations reveals that these two processes are not actually mutually exclusive. Interpreting foreign technology according to their own associations, the Japanese created new, original devices; and extrapolating from a series of foreign-born associations, they reached conclusions that were new and original *in Japan*, even if similar to European precedents. The process of interpreting new information according to one's own associations is a creative process. Radically put, there is no entirely original creation of knowledge or technology, only synthesis, and every new synthesis is original.

Why focus on astronomers? Although astronomers are often imagined as somewhat detached from society—both physically and conceptually—historians of science have pointed out that astronomers' techniques both are embedded in common practices and contribute directly to their transformation.[46] First of all, astronomers were the ones who determined the exact lengths of the variable hours during different seasons, and thus dictated the way everybody else in Japan measured time. Next, a key part of astronomical practice is the timing of celestial events, which made astronomers the most avid users of timepieces. They designed their own timepieces to suit their particular needs, thus crystallizing in matter their expectations of time-measuring processes. Astronomical timepieces therefore serve as a tangible manifestation of processes otherwise invisible, and provide an invaluable source for historians. Finally, a focus on astronomers is especially useful because, by the beginning of the Meiji period, astronomical time measurement came to be seen in Japan as superior to other forms of measuring time.

To focus on astronomical time measurement is not to assume that this mode of time measurement was somehow objectively superior, but rather to interrogate why late Tokugawa and Meiji intellectuals thought it was. They had not always held this opinion. The notion that astronomical time measurement was superior did not enter public discourse until several decades into the nineteenth century. Before that, even astronomers themselves did not necessarily think that the way they measured time at work was somehow better than the way they measured time at home, at a temple, or on a trip. It was simply different, as it was geared to a distinct purpose. Nevertheless, by the mid-nineteenth century numerous intellectuals—the ones who would later promote the calendrical reform—came to see astronomical time measurement as *the* model of time measurement. In an interesting historical twist, while the Europeans were slowly abandoning the idea that time represented an inherent aspect of physical reality, the Japanese, who had previously viewed time as a set of man-made conventions, now followed the older European way of thinking.[47]

The next several chapters focus on the spread of new perceptions of time measurement beyond a narrow circle of astronomers. Chapter 4 describes how astronomers' new approaches to thinking about the measurement of time made their way into the world of geography and surveying, and eventually manifested in a new map format. Focusing on the mapping project undertaken by Inō Tadataka (1745–1818) in the beginning of the nineteenth century, this chapter shows how astronomers' theoretical and mathematical descriptions of planet earth were translated into actual practices on the ground. Inō initially set out on his survey in an attempt to determine the exact length of one meridian degree at the latitude of Japan. This quest necessitated that he not only make terrestrial measurements but also corroborate them with celestial observations. Like the astronomers at the Asakusa observatory under whom he had studied, Inō used a pendulum clock to time celestial events and calculate distances trigonometrically. But while astronomers were drawing diagrams in which celestial bodies were connected to specific places on earth, Inō *was* that place on earth. By making time measurements along his surveying route, and by comparing measured times with those conducted in the Asakusa observatory, Inō created a link—a link of time—that connected various distant places in Japan. In so doing, he took an idea of the time-space relationship, previously confined to the theoretical and mathematical in astronomical practice, and implemented it on the ground.

This conceptual link created by the measurement of time then manifested in the maps that were produced on the basis of Inō's measurements. It is not a revelation that maps convey a variety of meanings.[48] Here, however, I would like to stress that those meanings are shaped not only by intention, scope, or conventions of depiction, but also—before anything was projected onto paper—by the very methods of surveying.

Measurements of time were translated into longitudes. And since, conventionally, Tokugawa astronomers used local time in Kyoto as the standard for calendars used throughout Japan, the central meridian—the zero longitude and thus "the beginning of time"—was set to Kyoto as well. The timing of celestial events alongside the terrestrial surveying translated into a map in which all the distant places in Japan were defined by and tied to the traditional cultural capital of Japan—Kyoto. This connection was no longer just social or political, it was inherent, defined by the celestial measurements.

Chapter 5 continues the theme of the spread of astronomical perceptions of time. It shows how one particular kind of time used by Japanese astronomers—Western time measured by chronometers—came to be seen as the *ultimate* kind of time. The chapter begins by describing how the need for astronomical data led to Tokugawa astronomers' use of British nautical almanacs, which calculated events according to Western conventions of time measurement. At that point, circa 1800, Western units of time were only one of the several systems of time measurement that Tokugawa astronomers used. Nautical almanacs, however, inspired contemplations of open-sea voyages—first as a mathematical riddle requiring the calculation of a route over a spherical surface, and then as a concrete aspiration to sail abroad. Extrapolating from the practices of establishing longitude on land, eminent scholar Honda Toshiaki proposed determining longitude at sea by means of time measurement. Realizing, however, that the preferred timepiece of astronomers at the time—the pendulum clock—would not work aboard a ship, he suggested using a Western chronometer. Use of the data in British nautical almanacs led Japanese astronomers and other scholars to the conclusion that the Western approach to time measurement was not just one among multiple possible conventions, but rather revealed a fundamental truth about reality in reflecting a mathematical average of earth's motion. If the maps of Inō Tadataka manifested the idea that time was not just a convention but was rooted in the structure of the universe, the developments surrounding the discourse about navigation propelled this idea one step further—now it appeared that Western time offered the most accurate representation of that structure.

Curiously, although much of the discussion of time measurement revolved around concerns about navigation, those concerns were mostly hypothetical. Though Tokugawa Japan was far from the "closed country" of the popular imagination, Japanese citizens were nonetheless forbidden from traveling abroad, and thus could not actually experience time measurement on the open sea.[49] That is not to say that Toshiaki's contemplations about navigation were pure intellectual musings disconnected from concrete experience. Quite the opposite is true. Toshiaki's ideas about time measurement at sea were based on very real practices of astronomical surveying on land. Using data from British nautical almanacs, determining longitudes on land, examining maps of the world, and attempting to measure time aboard the merchant ships that circled Japan—all these provided materials for contemplation, and for forming normative associations. Without being prompted by real-life necessities, people like Toshiaki could *extrapolate* from the practices they were familiar with and, combining elements they knew, actively form new ideals of time-measurement practice.

Chapter 6 explores the interplay of the various evolving notions of time measurement described in the preceding chapters by focusing on the work of one individual—Endō Takanori (1784–1864). Takanori was born into a high-ranking samurai family, in line to become an infrastructure minister of the Kaga domain, which roughly overlaps with present-day Kanazawa. In preparation for his vocation, he studied astronomy and surveying, and indirectly learned from another Kaga-born scholar: Honda Toshiaki. Takanori absorbed the notion of time as rooted in celestial motion and geographic location, and concluded that human measurement of time had to correspond with physical reality. Later in life he became interested in Western drawing methods and developed an aesthetic theory, according to which Western-style depiction offered a superior representation of physical reality because it was based on the same principles as astronomy and surveying. And finally, his interrogation of castaways who brought him a Russian clock led him to contemplate voyages abroad. Takanori's distinctive life trajectory was reflected in his evolving conceptualization of time measurement—as representative of physical reality, as associated with Western scientific practices, and as a necessary component of international travel. At the same time, these personal beliefs about the nature of time measurements clearly echoed the ones we encountered in the preceding chapter.

New conceptualizations of time measurement thus spread well beyond circles of astronomers. Chapter 7 focuses on the dissemination of ideas about time measurement by clock-makers. Clock-makers were the ones who produced measurement instruments for astronomers—not only custom-made astronomical clocks, but also a variety of compasses, theodolites, and sextants. While working on the designs of such instruments, clock-makers had to learn about astronomers' goals, practical needs, and models—an instrument brought by the Dutch, for instance, or one described in Jesuit books imported from China. Through this process, clock-makers also absorbed some of astronomers' perceptions of time measurement and time-measuring devices, which they in turn disseminated more broadly through the instruments they made for the general public.

Clock-makers were also the ones who explained to the public the principles on which Western clocks and watches were based. They built timepieces not just according to Japanese designs—they also took imported European timepieces, modified the mechanisms, affixed new dials, and made them usable in Tokugawa Japan. Adjusting the gears of European mechanisms to work in a different mode, clock-makers were the first ones to understand how to translate between the European convention of timekeeping and the Japanese one. Utilizing their knowledge, they then published treatises and pictorial guides to teach their readers how to use European clocks and understand the logic behind the time they displayed.

But why would anybody in Japan want to invest in learning how to read Western clocks? As chapter 8 describes, people made the effort because by the end of the Tokugawa period such clocks came to be associated with newly developed cultural ideals—science, progress, and internationalization.[50] Analyzing the rhetoric of the series of proposals to reform the calendar, I describe how reformers increasingly saw the Japanese calendrical system as "ignorant" and demanded a more "enlightened" one. As astronomy had come to be regarded as the epitome of rationality, astronomical timekeeping was now seen as a model. And since, by the mid-nineteenth century, astronomers were using Western marine chronometers, these devices themselves became identified with rationality.

With the advent of the Meiji period, astronomy and the related fields of geography and navigation became associated as well with the new ideals of modernization and Westernization. Thus, when people in Japan began evaluating the Western system of timekeeping as "better," they did so not because it was necessarily answering their practical needs more effectively, but because they identified it with a range of positive associations, such as "navigational use," "global commerce," and "astronomical tools." There was nothing in the Western timekeeping convention that better served the practical needs of everyday Japanese users than the system of variable hours. Rather, during that particular moment in Japanese history, the Western system of timekeeping was seen as superior because it was associated with practices that connoted emergent ideals of progress.

ONE

Variable Hours in a Changing Society

Hours

What is an hour? The answer to this question in Tokugawa Japan differed considerably from our answer today. First of all, there were only twelve hours in a day. They were referred to as either *toki, ji,* or *koku,* and written using a variety of characters.[1] In the English-language literature these hours are often dubbed "double hours," referring to the fact that there were half the modern number of hours in one day. The term "double hours," however, is fundamentally inaccurate, since Tokugawa-period hours were almost never 120 minutes long. This was due to the fact that hours changed their lengths throughout the year. The day was always divided into six hours of light and six hours of darkness. Because the relative lengths of days and nights changed with the seasons, the length of daytime and nighttime hours changed accordingly. If we translate Tokugawa scholars' calculations into our modern temporal units, daylight hours could be as short as 77 modern-day minutes or as long as 156 minutes.[2] Since most of this book deals with Tokugawa-period hours, we will refer to those by the term "hour," and distinguish our contemporary units by referring to them as "Western" or "modern-day" hours.

Of course, most of the people in Tokugawa Japan did not have any particular name for the temporal system that relied on variable hours—it was simply the way they measured time. From later in the eighteenth century, when Japanese scholars began comparing their timekeeping system to the one used by foreigners, they referred to the way they themselves kept time as *futeijihō*.[3] The literal translation of this term is "method of undetermined hours," which stresses that its hour-units were not bound by a predetermined length. However, as "undetermined" in English might imply a sense of hesitation and lack of resolution, we will refer to the temporal units of the *futeijihō* day as "variable" hours.

Each of the twelve hours was named according to one of the twelve "animal signs" and given a number that could be announced by striking a bell. Seeing that animal signs may be used for prognostication purposes, they are often misidentified as "zodiac signs" in English,[4] akin to the astrological signs of the Western world that derive from the constellations on the zodiac belt. Animal signs, however, were not related to Japanese zodiac constellations, but derived from an ancient Chinese cosmological system of twelve "terrestrial branches."[5] The "branches" were not originally identified with animals, and therefore, the signs are written with characters different from those used to signify the actual animals. [6] However, by the third century BC in China, the signs came to mean Rat, Ox, Tiger, Rabbit, Dragon, Snake, Horse, Sheep (or Goat), Monkey, Rooster, Dog, and Boar (appendix 1). Even today, traces of this system persist in the Japanese language—the terms for "a.m." and "p.m." in Japanese literally mean "before Horse" and "after Horse."[7]

Unlike the numbers on a Western clock, Japanese hours were not counted by consecutive numbers 1, 2, 3 . . . 12. Instead, the midnight hour (Rat) was hour number 9, and the following hours were numbers 8, 7, 6, 5, and finally 4, which was the final hour preceding noon. The noon hour was again numbered 9, followed by 8, 7, 6, 5, and 4, the hour before midnight. Thus, number 9 signified both noon and midnight—the ultimate points of daytime and nighttime. Number 6, on the other hand, referred to the points of change when day turns into night and vice versa. Accordingly, the morning hour of Rabbit was also called the Sixth of Dawn, and the evening hour of Rooster was also known as the Sixth of Dusk[8] (see fig. 1.1).

Wondering "Why do we count the hours the way we do?" several early eighteenth-century scholars investigated classical Chinese texts, and concluded that the origins of this double sequence could be found in the ancient *Classic of Changes* (the *Yi Jing*).[9] They came to believe that the system was supposed to represent correspondences between the twelve hours, the twelve months, and the annual cycle of birth and decay. In the *Classic of Changes*, each stage in the annual cycle is represented using a hexagram. The month of the year with the least daylight is when the dark *yin* energy is most dominant and the bright *yang* energy is just beginning its return. This month is represented by the hexagram called "return" ䷗, in which five broken lines symbolize the plentiful *yin* and the single unbroken one refers to the weak *yang*. During the next month, represented by the hexagram "approach" ䷒, the *yang* grows stronger, and continues to grow with each month until it

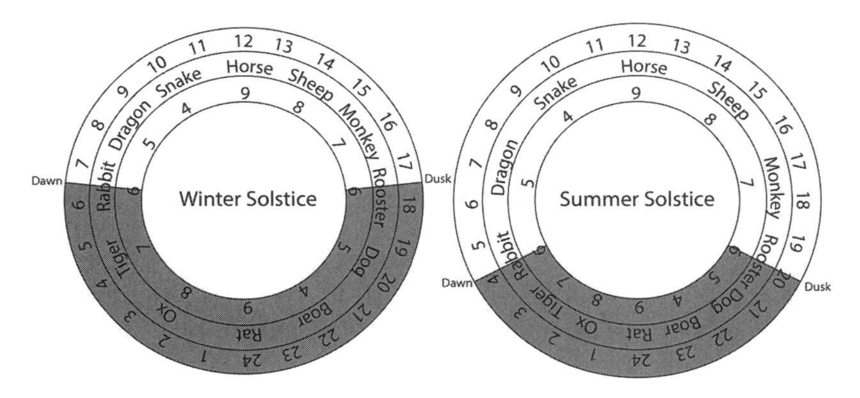

Figure 1.1. Schematic representation of the Tokugawa period system of variable hours. Diagram by author.
The two images show the difference in the lengths of variable hours from one extreme during the winter solstice to another during the summer solstice. The outer circles provide a comparative scale in hours used today. The middle circles show Japanese hours in animal signs, and the inner ones in numbers associated with those signs. The images are based on a calendrical algorithm developed for the 1798 Kansei calendrical reform.

reaches its peak in the month before the summer solstice, represented by the hexagram "(full) force" ☰. Since the cycle of hours mirrored the cycle of months, hours, too, were associated with the same series of hexagrams. Additionally, each hexagram—and hence each hour—could be discussed in numerical terms. Each unbroken *yang* line at the bottom of a hexagram was assigned the number 9. Multiplying the number of unbroken yang lines by nine gave the following series: the first hexagram was represented as the number 9 (1 × 9), the second as 18 (2 × 9), and so on: 27, 36, 45, 54. From this series, another series of numbers was derived by taking only the second digit of each number in the first series: 9, 8, 7, 6, 5, 4. Those were the numbers assigned to the hours from midnight to just before noon. Noon, in turn, corresponded to the summer solstice, in which the *yin* is born and begins its own process of growth. Since the two processes of growth—that of the *yang* and that of the *yin*—mirror each other, the hours from noon to midnight were also assigned the same series of numbers from 9 to 4.[10] It is nearly impossible to know whether these early eighteenth-century scholars were indeed right about the origins of the number sequence, nor did they themselves claim their conclusion to be an undeniable truth. They were part of a broader Tokugawa intellectual movement of *evidential research* (*kōshōgaku*) that relied on a philological approach in scrutinizing ancient Chinese texts.[11] As such, they were looking not for metaphysical truths but for historical evidence to explain existing practices.

What Tokugawa-period scholars agreed on was that their temporal system was human-made and used in Japan out of convention.[12] According to major sources from the period, whatever the exact origin of the 9-to-4 sequence was, these numbers were used in Japan by virtue of a conscious human decision.[13] These sources agree that in the seventh century, Japanese emperor Tenchi decided to adopt the Chinese timekeeping system, which at the time used variable hours counted in the double 9-to-4 sequence.[14] Following his decision, Japanese society used this method of hour counting for more than a millennium.

Although there were a variety of technologies available to measure hours of variable length, the most common device employed for this purpose was the incense clock—*jikōban*[15] (see fig. 1.2). An incense clock was a wooden box filled with sand. Using a template, one would create a channel in the sand and then fill it with powdered incense. Alongside the incense, the user would arrange hour indicators—which looked like little nails—at appropriate distances as signposts to indicate the hours. By watching the progression of the burning incense relative to the indicators one could know the time. Since the indicators were not fixed, it was possible to place them at varying distances according to the season. It was also possible to fill the channel with incense that produced different aromas, so that the change of hours could be detected through the sense of smell.

The use of incense clocks, as well as other nonmechanical timekeepers, did not disappear with the importation and gradual proliferation of mechanical clocks. We find a wonderful description of an incense clock by the Russian naval captain Vasiliy Golovnin, who was captive in Japan from 1811 to 1813 and observed the operation of such a device, at the time still in common use:

> [a] relatively small wooden block, covered with clay and whitened; in the clay a narrow channel is drawn, and filled with powder made of some kind of grass, which burns very slowly, and on the sides of this channel there are holes into which one inserts a nail; near those holes there is a designation of the length of day and night hours during the six months from the vernal to autumnal equinox; during the other six months, the day hours become the night ones and vice versa. Thus, the Japanese clock masters find the length of the daytime hour in a certain period, mark it with the nail and, filling the channel with powder, they set it alight at noon, and in this way they measure time. This wooden block they keep in a closed box and try to store it in a dry place.[16]

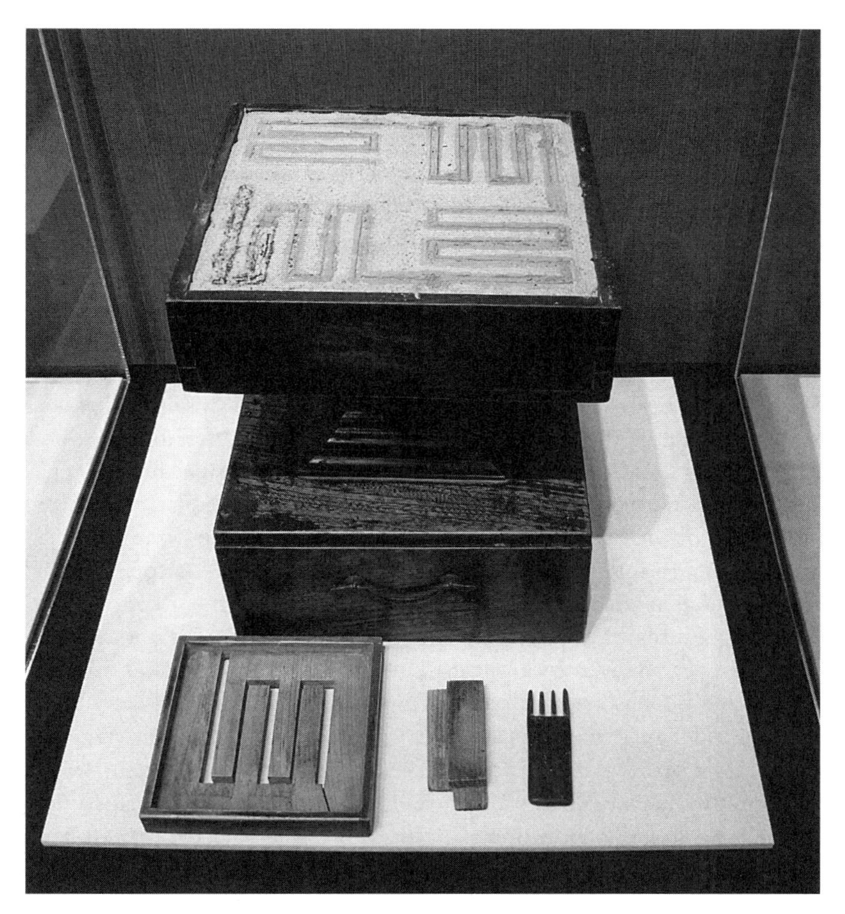

Figure 1.2. Incense clock. The Seiko Museum, Tokyo.
The keeper of the incense clock would first smooth the ashes on top of the device and then use the template (lower left) to tightly pack incense powder into a geometric pattern. He would then arrange indicators (missing here) along the incense path at intervals corresponding to the lengths of hours during the given season. Once ignited, the progression of burning incense relative to the indicators revealed the hour of the day or night.

Seasons

But how did one determine the length of the hours? The system of variable hours might evoke nostalgic notions of "natural" time, attuned to the motion of the sun and the seasonal changes in the length of the day. One might even assume that daily changes in the length of the hours were simply deducted from natural phenomena. Looking closely at the calculations required to maintain such a system, however, we see that it was anything but "natural." First of all, the system of variable hours required a definition of the exact moment that separated night and day. There is no "natural" way to determine this moment. Intuitively, it is very tempting to claim that day begins when the sun appears above the horizon and ends when it disappears below it. But, to mention just one of the many problems with this approach, unless one is at sea, it is difficult to define what the "horizon" exactly is. More importantly, the sky begins to get light long before the sun itself rises above the horizon, and what we refer to as "dawn" is not a specific point but a gradual process that sometimes takes more than an hour. Consequently, in terms of brightness, the day begins at some point before the sunrise. There was an old saying that advised one to look at one's hand: day has arrived if there is enough light to distinguish the three main lines on the palm. This is an obviously subjective measure, dependent on one's surroundings, the weather, and even one's eyesight. For a society to function, there had to be an agreement as to which moment counted as the beginning of daytime.

The beginning of daytime, and hence the length of hours, was defined with reference to an astronomical timekeeping system of equal units. We will discuss how astronomers decided on the parameters for determining the beginning of daytime in chapter 3, but for now let us focus on the temporal units they used before the nineteenth century. In astronomical practice, the day was divided into one hundred equal units called *koku*, each a little more than fourteen minutes in modern terms.[17] If all hours were of equal length then each hour would have been eight plus one third of a *koku*, but since hours changed their length with the seasons they were rarely that length. Astronomers determined at what *koku* the day began and ended. Dividing that interval into six equal parts they then arrived at the length of daytime and nighttime hours for a particular season.

The name of these units—*koku*—derived from an ancient timekeeping device, a five-vessel clepsydra—a *rōkoku*—that was introduced to Japan from China together with the timekeeping system[18] (see fig. 1.3). Dripping from one vessel to another, there was a steady flow of water that gradually filled the bottom vessel. There, a figurine of a man with an arrow in his hand floated. The tail of the arrow was marked with one-hundred notches. As water dripped into the vessel, the floating figurine rose. By observing how many notches on the arrow's tail were visible above the top edge of the vessel, it was possible to know how much time had passed since the device

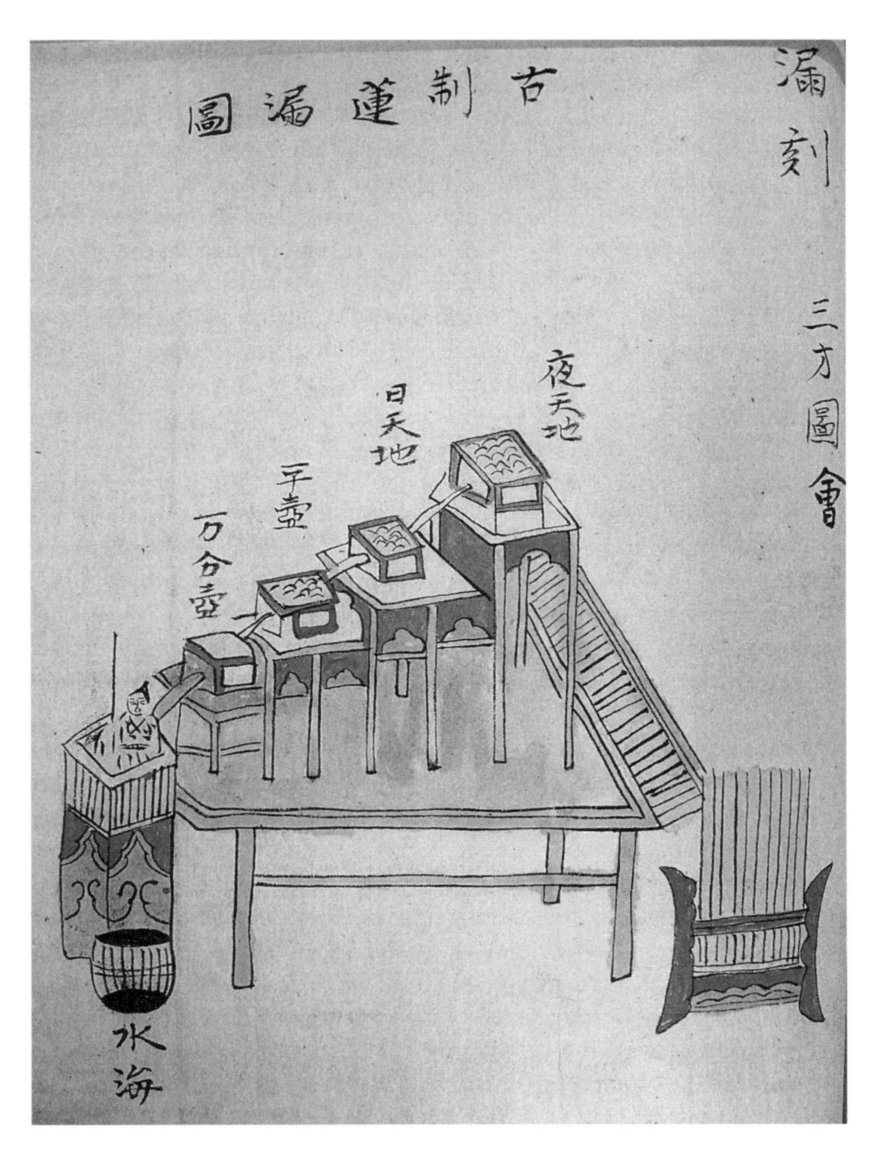

Figure 1.3. Clepsydra. *Illustrations of Various Measuring Devices* (*Sokuryō sho ki no zu*). National Astronomical Observatory of Japan, Mitaka.
A sketch of the ancient clepsydra by eighteenth-century astronomer Nishimura Tōsato. Water drips from one vessel to another and accumulates in the bottom vessel, which raises the floating figurine holding an arrow marked with notches that indicate the passage of time. The dripping of the water and the notches on the arrow supply the name of the device—*rōkoku* (漏 dripping + 刻 notches). The latter character is still used in modern Japanese in some words related to time.

was calibrated. Reflecting its crucial physical characteristics, the name of the device comprises the two characters "dripping" (*rō* 漏) and "notches" (*koku* 刻). The hundred equal-length "notches," or *koku*, thus came to signify the basic astronomical division of the day into equal temporal units.

The length of hours was not adjusted on a daily basis. The daily change in the amount of daylight is almost undetectable, and did not amount to even one *koku*. Nor was the rate of change steady throughout the year, so one could not simply change the length of hours after a predetermined, consistent interval of days in which the daily change in the amount of light amounted to one *koku*. Instead, people adjusted the length of hours twenty-four times a year according to seasonal units called *sekki*,[19] each of which measured approximately fifteen days.[20] Each *sekki* received a name signifying seasonal changes either in atmospheric phenomena or related agricultural activities, such as "*great heat*," "*white frost*," "*rain on the grain*," or "*beginning of spring*" (see appendix 2). Other *sekki* names indicated the beginning of one of the "major" four seasons,[21] or important astronomical events such as equinoxes and solstices. Each *sekki* was further divided into three environment-oriented episodes called *kō*,[22] which were supposed to represent more specific changes in flora and fauna, such as the appearance of insects after their winter hibernation, the return of migrating birds, or the blooming of various flowers and trees.[23]

Looking at the names of the *sekki* and *kō*, it is very tempting to conclude that Tokugawa Japanese were highly attuned to minute seasonal changes and the natural phenomena around them. They may indeed have been attuned to seasonal changes, but this is not something we can deduce from the terms of *sekki* and *kō*. The names used in Tokugawa Japan were inherited from the ancient Chinese calendar and therefore, if anything, were supposed to reflect changes in the natural environment of northern China, where the system was created.[24] Consequently, we should not be surprised to find that the list of *kō* featured animals like tigers, which do not inhabit the Japanese islands, and that the timing of natural phenomena according to *kō* did not coincide with when they actually occurred in Japan. Not to mention the fact that Japan itself is far from climatically homogenous, which made those names even less indicative of natural cycles in various regions.

People, of course, were not oblivious to these discrepancies. After all, it is not hard to notice that the period called "*beginning of Autumn*" was the hottest period of the year, while the period "*snow turns into rain*" could in fact bring sudden snowstorms to areas that do not see much snow during the winter at all.[25] Scholars had already begun to problematize this discrepancy in the seventeenth century, urging the adoption of a different list of *kō* names that would better correspond to natural events in the Japanese climate.[26] Scholarly debates notwithstanding, the general population was reluctant to part with convention, and kept using the less accurate yet familiar Chinese *kō* names.

Calendars

There was another incongruity that simply could not be ignored—the incompatibility of solar seasons with lunar cycles. As in many other societies, a month in Tokugawa Japan was defined by the cycle of the moon. But that cycle is 29.53 days long, and therefore cannot be defined in terms of whole numbers of days.[27] To adjust this odd length to practical standards, shorter and longer months were defined, respectively comprising twenty-nine and thirty days. But which months should be short and which long? In our modern-day calendar, short and long months alternate (most of the time). But if one wants to keep months aligned with the lunar cycles, this solution is imprecise. Furthermore, twelve lunar months add up to only 354 days, so if one wants the same months to occur in the same season year after year, a periodic intercalary month is required.[28]

Thus, not only did the system of variable hours—and the luni-solar calendar it relied on—require human regulation, but it required *more* regulation than the solar calendar we currently use. Since the luni-solar calendar had to adhere to a multitude of requirements, calendrical patterns were based on multiyear cycles. The same sequence of long and short months repeated only every forty-three years. The solar-based *sekki* repeated their position relative to the twelve lunar months only every nine years. And the position of the intercalary month was based on a period of forty-seven years.[29]

As it was impossible to keep all these variables in mind, people had to rely on paper calendars, produced and distributed by the government.[30] By the beginning of the Tokugawa period, a standard calendar would have contained all the information necessary to coordinate social activities.[31] First, the calendar indicated the year according to the *nengō* system of eras, lengths of time marked by certain important events like changes in imperial succession, important political reforms, or major earthquakes[32] (appendix 3). Immediately after the *nengō*, the calendar indicated the year according to a sexagesimal cycle, the *kanshi* or *eto*.[33] This was a one-to-sixty count created by a combination of what is usually referred today as the ten "celestial stems," *kan*, and the twelve "terrestrial branches," *shi*, also known as the animal signs (appendix 4).

The calendar indicated how many days a particular year had, and which months were long (thirty days) and short (twenty-nine days). It offered information about astronomical phenomena such as solstices, equinoxes, and eclipses, as well as about social events such as holidays and festivals. Finally, it included a large middle section of "calendrical pointers" that instructed people how to manage their daily activities.[34] These calendrical pointers relied on prediction—*uranai*. [35] *Uranai* was a general category that included meteorological guidelines, such as the end of the frost hazard, the beginning of the monsoons, or the start of the typhoon season. It also included predictions based on hemerology — the art of prognostication based on calendrical calculations. Hemerological pointers indicated auspicious and inauspicious days, favorable directions, and/or instructions for specific activities such as building a house or getting married.[36] (see fig. 1.4).

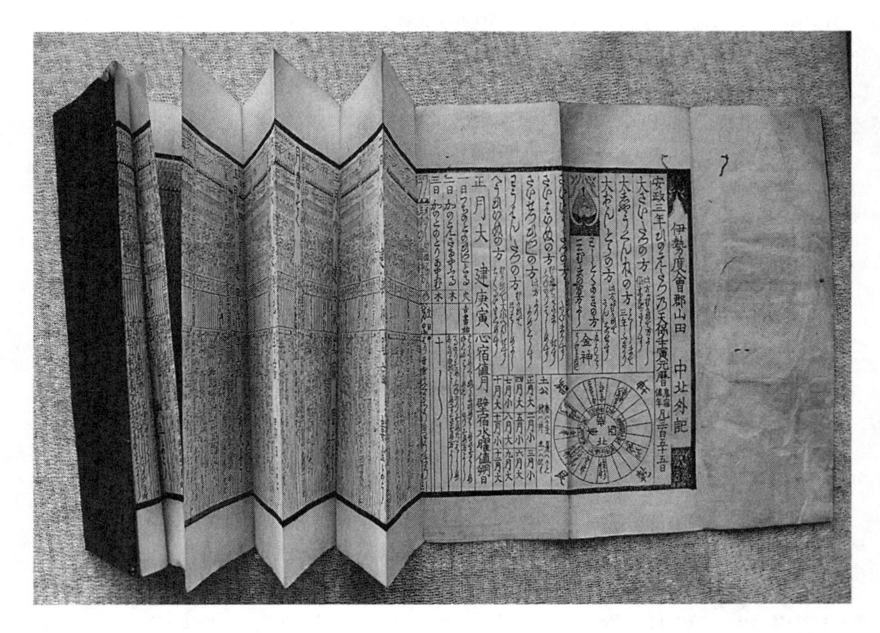

Figure 1.4. Calendar from the year Ansei 3 (1856). Author's private collection. The calendar is presented as a chart that provides information for every day of the year. In addition to indicating the twenty-four seasons, the long and short months, holidays and festivals, and climatic periods such as the monsoon, in its middle section the calendar also provides hemerological information about the auspiciousness of certain days. The calendar was folded into a portable booklet that could be carried around and opened to the appropriate panel.

Paper calendars were indispensable. As the author of *Explanation of Calendrical Pointers* (*Rekijitsu chūkai*) wrote:

> A calendar takes the succession of the sun and the moon, and, based on that, determines the four seasons, decides on *sekki*, and declares lengths of the day and the night. . . . [The calendar] shows the weather, drought, calculates the time of tilling and sowing, and therefore it is the most valuable asset of human beings. Consequently, whatever they do, first of all people open the calendar.[37]

It is unsurprising, then, that the compilation of the calendar was considered to be one of the core responsibilities of the government. The regulation of time and calendrical calculations began immediately after the introduction of the temporal system and related cosmology from China in the seventh century. An astronomical bureau was established shortly after, composed of four divisions: timekeeping, astronomical observation, calendar compilation, and hemerological interpretation.[38]

The beginning of the Tokugawa period saw a sharp increase in the number of calendars produced, due both to the rapid development of print culture and to the appearance of new sites of calendar compilation. The proliferation of local calendars in the seventeenth century caused quite a lot of confusion. Generally speaking, the system described above was supposed to be employed in each of the calendar production sites. But in reality the complexity of astronomical and astrological calculations, geographical differences, and the inevitable arbitrariness of some of the decisions made during the process, as well as local variation in the parameters that were taken into consideration, resulted in divergent calendars.

In 1685, citing an urgent need for a standardized calendar, the shogunate carried out a calendrical reform and began requiring the domains—or, more colloquially, provinces; that is, the various territories controlled by Japan's *daimyōs*, or local lords—to abide by a calendrical template created by shogunal astronomers. Failure to follow the central template or intentional alteration of the template could result in the revocation of one's license. This legislation smoothed communication and commerce between the domains, and gave the central government more control over them through the control of time. Since the length of seasonal hours was determined by calendar, this effectively meant that the government controlled and regulated the hours in most of Japan.[39]

The annual calendars, sponsored and distributed by the central government, were only one type of calendar used by the general populace. These full-length calendars, which detailed information for each and every day of the year, though often bound or folded into a small book, were still too long and cumbersome to be consulted for basic information like the length of the current month. As a more convenient solution, one could use an abbreviated version of the calendar (*ryaku-goyomi*), which was only one page long and compactly presented only the most important information—which months were long or short, when the intercalary month fell, when the rainy season began.[40] An even more concise version was the "big-and-small" calendar, which showed only which of the months were short and long. [41]

Another curious type of calendar could be found in the northeast. Quite remote from the main urban centers by early-modern standards, this region was considered to be wild, exotic, and intriguing by "sophisticated" Edoites, and it was not uncommon for intellectuals to take long trips there in order to connect with nature and explore the culture and tradition of the "savage" northerners, in the same way they had explored other "exotic" cultures such as the Chinese or Dutch. One of those explorers was Tachibana Nankei, a popular writer of the end of the eighteenth century, who had discovered that people in the northern city of Morioka used pictorial rather than written calendars.[42] Instead of writing down the lengths of the months, the dates of important festivals, and weather predictions, all this information was encoded in pictures. Edoites referred to those calendars as "blind calendars," suggesting that northerners were forced to use pictures because they were blind to civilized ways, namely illiterate. (see fig. 1.5).

As Tachibana himself noted, deciphering pictorial information was not an easy task at all, and certainly required a specific kind of pictorial literacy, which was not necessarily less complicated than the basic alphabetical *kana* style in which regular calendars were written. In addition, the pictorial style was not foreign to central Japan either, where one could readily

Figure 1.5. A "Blind Calendar," 1826. National Diet Library Digital Collections. The top row of the calendar indicates the era and the year, encoded in pictographic word play that requires at least some degree of literacy to decipher. On the right we see a picture of a letter (*bun* in early modern Japanese). To the left of the letter is a man carrying something on his back (*se*); the thing he carries is the character for "well" 井 (*i*). The combination of the three images gives us the name of the era: Bunsei (文政). Next to it we see a dog, the animal sign of the ninth year of that era, also indicated by the numbers on the dice—one plus three plus five. The column on the left, beneath the short sword, shows the short months, while the column on the right, beneath the long sword, gives the long (e.g., 30-day) months. The images beneath indicate festivals and seasonal events, such as the beginning of the monsoon (indicated by a thief carrying away goods) or midsummer (indicated by an old man wiping sweat from his forehead).

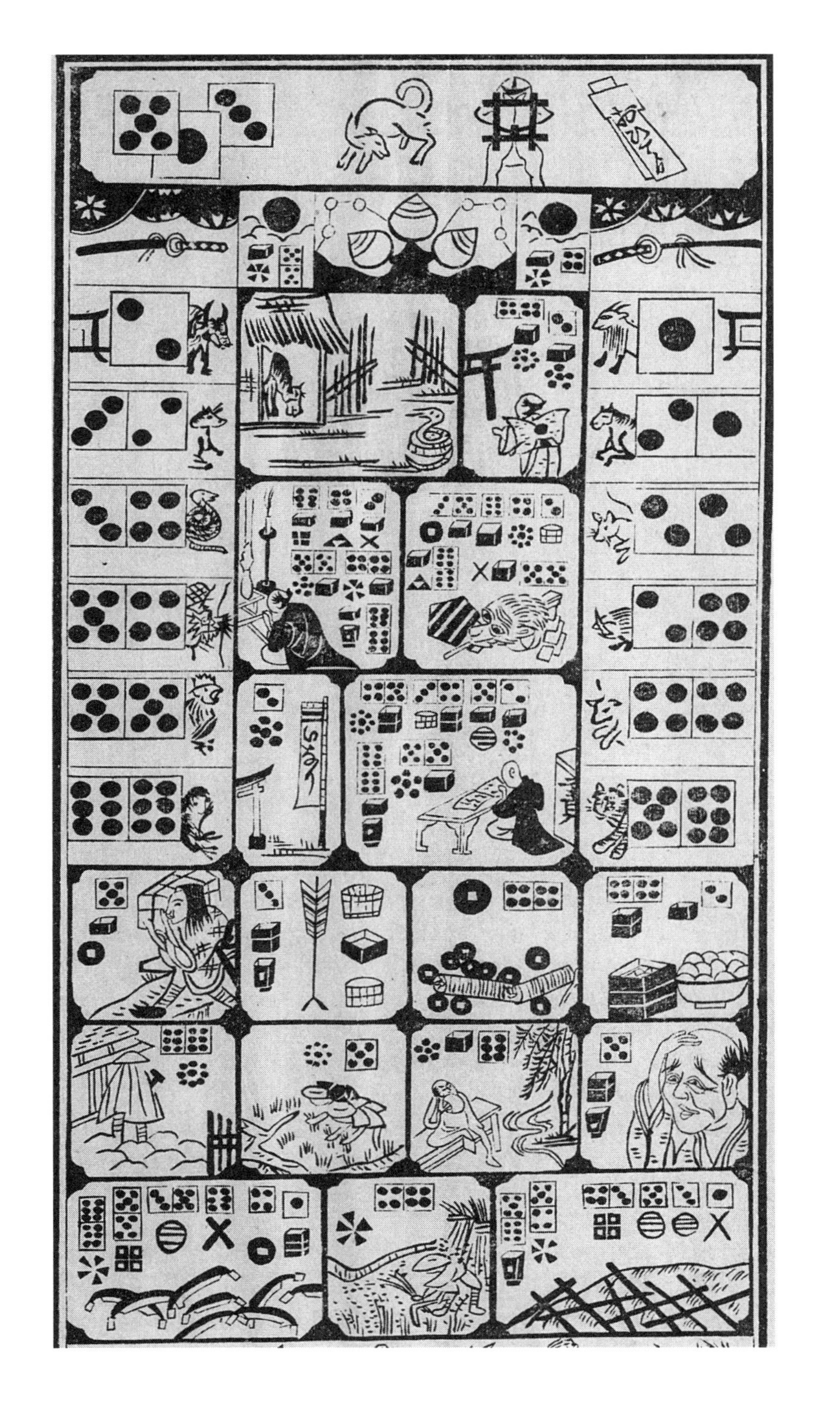

find abridged calendars rendered in pictures. But more importantly, many names were encoded in word-, character-, and sound-play, so that to understand the pictorial hint, one had to have some knowledge of written characters. Consequently, it is much more likely that the pictorial style of the northern provinces was a matter of local convention rather than an accommodation of illiteracy.[43]

So how many calendars, of all kinds, were out there? According to Watanabe Toshio, a historian of astronomy and calendrical sciences, there were enough for every household to receive at least one calendar each year.[44] Along the same lines, Okada Yoshirō, who conducted extensive research on calendar use, argues that in the late Tokugawa period around four and a half million copies of official calendars were published each year. With a population of around twenty-seven million, this meant that there was about one official calendar distributed for every four people (including small children).[45] Neither scholar includes the abbreviated calendars, pictorial calendars, local newsboards that published full or partial calendars, or all the pirated, illegally sold versions. Indeed, even the risk of severe punishment did not deter those who identified an economic opportunity in calendar-making, answering the needs of a market with high demand. Taking into account all of the above, it is safe to say that many people probably owned several calendars, some of which were specifically designed to be carried around.

Time-Bells

The standardization of calendar production near the end of the seventeenth century also affected the regulation of daily time measurement. When the seventh-century Japanese rulers adopted the Chinese astronomical system, they built an astronomical bureau in what was then the capital—Heian, or Kyoto, as it later became known. Since most astronomical observations were conducted there, astronomers used Kyoto time as their standard. The practice of using Kyoto time as the calendrical standard continued in the following centuries. Even when the government moved to Edo (Tokyo) in the early seventeenth century, the calendar was still calculated on the basis of the local time in Kyoto. Since calendars also indicated the length of hours, this meant that across Japan hours were now determined according to Kyoto time.

With the standardization of the calendar in the Tokugawa period, Japan officially had one time zone, but it must be stressed that this unification of time was in theory only. There was no technological means to assure synchronization across different regions of the country. Timekeeping devices in different regions were calibrated according to the local noon. Thus, even if astronomical calculations were made on the basis of Kyoto time, the actual measurement of time was dependent on the geographical specifics of each region. Still, we should not underestimate the importance of the ideal of standardized time, even if it was not achieved in practice.

Even without the means to achieve countrywide synchronization, there were attempts to synchronize the measurement of time on a local level through a synchronization of the announcement of time. In Tokugawa Japan, time was announced using sound. In the beginning of the Tokugawa period, the sound of time mostly emanated from temple bells, or from castles where the time was announced by beating a large Taiko drum. Once the city of Edo became the seat of the Tokugawa shogunate, it grew rapidly. At the beginning of the seventeenth century it was home to about 20,000 people; by the end of the century, its population was over a million. As Edo grew, people began complaining that they could not hear the bells anymore, and demanded that the government erect additional time-bells, or *toki no kane*[46] (see fig. 1.6). The project of erecting numerous bells was made possible by the booming economy of the end of the seventeenth century. The flourishing mining industry—fueled by the Dutch demand for copper—increased the availability of metals, so that not only Edo, but also smaller communities could afford to install bells.[47] One can often spot *toki no kane* in old maps and paintings, and a few are still extant. Golovnin, the Russian naval captain, described the time-bell he heard in the northern province of Matsumae:

> The Japanese strike the hours in the following manner: first they strike the bell once, then after about a minute and a half they strike twice, one strike right after the other; those three strikes announce that the hours are about to be struck, as if they were saying: listen! Then, after another minute and a half they start to strike the hours, strike after strike in intervals of some fifteen seconds, but the last two they strike quickly one after the other, as if to indicate: enough counting![48]

The ubiquity of time-bells prompted attempts to synchronize their sound. *Toki no kane* were installed in towers that allowed their sound to be carried over long distances, a defining element of the Tokugawa soundscape. By the beginning of the eighteenth century there was no place in the urban areas where time-bells could not be heard.[49] In fact, their number and proximity meant that their rings eventually began to overlap with each

Figure 1.6. The Asakusa time bell (*toki no kane*) depicted in the *Forty Eight Views of Tokyo* (*Tōkyō meisho yonjū hachi kei*). Shōsai Ikkei, Print on Canvas, early Meiji Period. Waseda University Kotenseki Sogo Database of Chinese and Japanese Classics. The Asakusa bell is one of the earliest time bells, and also one of the few that survives. The bell was famously immortalized by Matsuo Bashō in a haiku: "A cloud of flower blossoms; is the bell from Ueno or from Asakusa."

other. To eliminate confusion, bell keepers had to learn to synchronize their ringing with that of their nearest neighbors.

Although we usually associate synchronization with advanced mechanical technology, in seventeenth-century Japan it was nonmechanical timepieces that proved to be the more effective tool to achieve at least a certain degree of synchronization. Bell keepers rang their bells according to timetables, which indicated at what *koku* one should ring the bell during different seasons. In order to know what *koku* it was, bell keepers kept several timekeeping devices—both nonmechanical and mechanical.[50] Incense clocks proved to be surprisingly suitable for the purpose of synchronization. Although the rate at which incense burns, and hence its measurement of time, is influenced by climate, all devices in the same climatic area were affected similarly. Consequently, even if their measurement of time was incorrect in astronomical terms, they all deviated from the standard in a similar manner. Mechanical clocks, on the other hand, broke, sped up, or ran late in their own individual and idiosyncratic ways, and hence were often of no help to those who aimed to synchronize the ringing of time-bells.

Time Consciousness and Variable Hours

One of the reasons there was such high demand for time-bells was the practice of constantly noting the time of the day. Already at the end of the sixteenth century—even prior, that is, to the mass installation of time-bells—a Jesuit priest, Jean Crasset, noted that "the Nipponese have a great need to mark in their history, not only the day on which events occurred, but even the hour and the part of the hour."[51] This tendency only increased with the spread of time-bells. When the famous haiku poet Matsuo Bashō embarked on a journey to the north in the late 1680s, he traveled with a companion named Sora, who kept a detailed diary.[52] While Bashō was composing haiku, Sora was documenting everything that happened during the day: their arrivals and departures from inns and temples, lunches, earthquakes, the beginnings and ends of rains—all were described in detail. Almost every entry has a time stamp to half a variable hour.[53] The practice of timing even quotidian events was thus an important part of one's daily routine, and the bells enabled one to know the hour even in the remote "interior" of the country. Sora's diary was not at all exceptional. The practice of indicating time in one's diary is also vivid in the story of Yūyama Gin'emon, who lived later in the Tokugawa period. Yūyama was involved in a series of litigations and he kept a diary describing his ordeals. He usually recorded the hour when he left his house in the morning, when he arrived at the courthouse, and when he came back home. And he did not stop there. He recorded how long he was kept waiting at the court, how much time he spent eating lunch, when he decided to go for drinks, when he got a post-drink snack, and when on occasion, late at night, he decided that he needed another drink. Even in an uncertain state of consciousness, at a late hour, while trying to relax, Yūyama felt the need to note the time.[54]

Not only leisure, but work, too, was defined by hours. Although it was common to say that people worked "six to six"—that is, from the sixth hour of dawn to the sixth hour of dusk, in reality people were very much aware of the fact that "six to six" in the winter meant a different amount of time than during the summer. In general, people were paid by the day, yet this length of the "workday" was defined by contract and not by the cycle of sunrise and sunset. Contracts were adapted to the seasons, so that during the summer people would leave for work around hour 5 (Dragon) of daytime—an hour *after* the day begins with the dawn hour 6 (Rabbit). During winter, however, they would begin work already in hour 7 (Tiger) at nighttime, which was one Tokugawa hour *before* the dawn. (See appendix 1 for details about the countdown of hours). These adjustments resulted in a fairly even amount of daily work time throughout the year.[55]

As the Tokugawa period progressed, the indication of work hours became increasingly detailed. A collection of legal documents from first half of the nineteenth century concerns smiths whose contracts strictly defined the working hours in each season. In order to discipline tardy workers, something not unlike a punch-card system was used, in which people had to stamp their personal seal on a form at the beginning and end of every workday in order to receive their pay.[56] The contracts did not serve the employers alone. They also stated that the workers were entitled to a day off on every first, eighth, and fifteenth of each month, as well as during holidays and festivals.[57] Moreover, some of the contracts indicated that if a worker had worked more than what was defined as a working day he would be eligible for an overtime wage, calculated on an hourly basis plus 10 percent for every additional hour.[58] Times of work, rest, meals, and leisure—all were structured by the hours announced by the bells. As we shall see in the next chapter, this increasing time consciousness was also related to the growing number of timepieces, nonmechanical and mechanical alike.

* * *

The system of variable hours was not a simple matter of adhering to the daily and yearly cycles of the sun, but rather an elaborate set of temporal conventions. It had to be regulated and maintained, and was therefore far from "natural." The temporal categories it relied on persisted through millennia, survived numerous social and cultural upheavals, and as such did not reflect specific social structures. These conventions effectively shaped some of the basic structures around which social practices were built, even when they were no longer connected with their original meanings. They are akin to the foundations that determine the overall layout of a house but have no power to dictate what life its inhabitants lead. Take, for instance, another set of conventions—the use of length, weight, and temperature measures in the United States. Non-Americans consider the metric and the centigrade systems to be rational and convenient, and are usually taken aback by America's "backward" miles, ounces, and Fahrenheit degrees. Yet the use of these "old" units of measurement does not prevent Americans from enjoying modern lifestyles. Nor did twentieth-century modernization require getting rid of metaphoric, nonscientific units of measurement (e.g., "feet"). In the same way that the use of nonmetric units says nothing about the social structures of twenty-first-century American society, the variable hours of Tokugawa-period Japan were not a reflection of the specific cultural characteristics of Tokugawa society.

Since Japan eventually adopted the Western timekeeping system, there is a tendency to assume that the earlier temporal order in Japan was somewhat more lax than the one that arrived from the West. In fact, and despite the "natural" associations Tokugawa-period temporal categories evoke, the system of variable hours required more regulation and maintenance than the European temporal system. The sheer number of factors that needed to be considered in the calculation of hours required reliance on an organized system. One could not just replicate the same practice day after day, year after year, but had to rely on calendars. Already by the end of the seventeenth century, the Japanese government had standardized the calendar and systematized the announcement of the time using bells. The proximity of time-bells in dense urban population centers created a problem of overlapping announcements of the hours that called for synchronization.

Surprisingly, sometimes it was precisely the factors that *seem* to be incompatible with modern lifestyles that encouraged and enabled increased synchronization and time-consciousness. We may think of enormous public clock-faces as a sign of modernity, but it was auditory signals that made clear the inconsistencies between overlapping bells. We may think that mechanical clocks are more reliable than burning incense, but in the seventeenth century incense clocks proved to be more conducive to synchronization.

Of course, punctuality and synchronicity are relative terms. Needless to say, during most of the Tokugawa period people did not carry watches that measured minutes and seconds, and did not have to catch trains leaving on the minute. Yet this does not mean that they were not punctual—only that they were not punctual by *our* standards. To be punctual is to abide by a range of temporal approximations determined by existing social norms. Even in our modern society with our atomic clocks we always resort to approximations—how late must one be in order to be considered late? One millisecond? Three seconds? Thirty? Three minutes? Or maybe fifteen? The answer depends on the situation, not on the units of measurement.

As life during the Tokugawa period changed, people adjusted their temporal practices and their sense of punctuality—all within the conventions governing the system of variable hours. Instead of changing those conventions each time a new social practice emerged, they adjusted their practices within the existing system of timekeeping. And, as we shall see in the next chapter, they also found a variety of ways to adjust timepieces to their emerging needs.

TWO

Towers, Pillows, and Graphs: Variation in Clock Design

Wadokei

The year 1551 was the first time a mechanical clock was seen on Japanese soil. Brought by the Jesuit Francis Xavier, it was presented as a gift to Ōuchi Yoshitaka in gratitude for allowing the opening of a Christian mission in his province.[1] News of the fascinating object spread, and the Jesuits reported that they were asked more and more about the device that could strike the hours without the touch of a human hand. A decade later, the most powerful warlord in Japan, Oda Nobunaga, an admirer of everything Western, sent for another Jesuit, Louis Frois, demanding to see a clock—an "alarum."[2] His curiosity satisfied, however, Nobunaga said that "although he liked it, he did not want it because it would be useless in his hands."[3] Reading Frois's description of this conversation, it would be easy to assume that the Japanese warlord simply did not know how to handle the complex mechanical device. But the truth is that, from the Japanese perspective, European clocks were useless, since they measured time in equal units that failed to reflect the seasonal changes in the amount of daylight.[4]

Despite this, interest in European clocks prompted engagement with their mechanisms, which eventually resulted in the adaptation of those mechansims to the Japanese timekeeping system. When a clock presented to the first Tokugawa shogun, Ieyasu, broke, he entrusted the repairs to local ironsmith Tsuda Sukezaemon, and subsequently hired him to work exclusively on clocks.[5] Tsuda taught the craft to his children, thus establishing a hereditary line of clock-makers (all of whom were named Tsuda Sukezaemon), and initiating the profession of mechanical clock-makers in Japan.[6] As Tsuda was working on the clock, attitudes toward Europeans were already tainted with suspicion. Within a few years, the shogunate restricted the European presence to little alcoves—first Hirado, and then the artificial island of Dejima, which, as described in the introduction, was built off the coast of Nagasaki specifically for the purpose of hosting Dutch merchants at a distance from the Japanese lands. Aspiring Japanese clockmakers had no means of learning the craft from Europeans, and thus were left to interpret the foreign mechanism on their own. Drawing from their previous experience with nonmechanical timekeeping devices, they transformed European clock mechanisms to create a set of objects retrospectively called "Japanese clocks," or *wadokei*.[7]

The transformation of European clocks in Japan began even before any physical modification of actual clockwork. European clocks were recognized as serving a function analogous to that of existing Japanese timepieces—*tokei*. The modern-day characters for this word are 時計, meaning "time/hour" and "gauge/measure." Before the nineteenth century, however, this was just one of the possible ways to write the word *tokei*. Other character combinations reveal associations with numerous nonmechanical means of gauging the passage of time. Examining Tokugawa-period sources, one sees combinations such as 士景—"earth-shadow," 斗計—"measuring the Big Dipper," 士卦—"earth–trigram" (referring to the *Classic of Changes*, the *Yi Jing*), or 士圭—"earth-gnomon."[8] All these combinations of characters read as *tokei*, suggesting that the word was associated with a variety of concrete methods of time measurement—shadows, gnomons, stars. Dubbed *tokei*, European clocks came to be recognized as belonging to the category of timekeeping devices.

Nonmechanical Timekeepers and Mechanical Clockwork

Once European clocks were identified as *tokei*, they were interpreted through the lens of all the other kinds of *tokei* that already existed in Japan. Thus, for example, immediately upon their arrival in the country, European clocks were referred to as "self-sounding bells," or *jimeishō*—a nickname that stuck for the rest of the Tokugawa period.[9] Seeing this term, it is tempting to conclude that it reflected the astonishment of Japanese spectators at the device's mysterious ability to sound a bell without the touch of human hand. In reality, however, the term was not at all a neologism, and the phenomenon of sounding the time was not at all unprecedented. When Japanese spectators witnessed European clocks chime at a predetermined time, they identified the alarm function with an alarm-sounding device they already knew—the clepsydra, the water-clock discussed in the preceding chapter. In addition to measuring the passage of time using the steady flow of water, the clepsydra, or *rōkoku*, could also sound an alarm when the water in the bottom vessel reached a certain level. Due to this ability, and following the existing Chinese terminology, the device was called a *jimeishō* as early as the Heian period (794–1185).[10] When European devices were seen performing the same function, they, too, were dubbed *jimeishō*.[11] Later in the Tokugawa period, the term came to be used exclusively for mechanical clocks. That usage was thus based not on an astonished reaction to a mysterious ability, but rather on the association of the alarm function with a device with which sixteenth-century Japanese spectators were already familiar.

A similar process of understanding foreign technology by identifying it with something familiar took place when aspiring Japanese clock-makers sought to make sense of European clock mechanisms. The first ones that Japanese clock-makers encountered were of a type dating back to the fourteenth century, in which the clock was driven by the power of falling weights and controlled by a verge-and-foliot escapement (see fig. 2.1). The verge is a vertical rod with two pallets that hold and release the teeth of the crown wheel. The foliot is a horizontal balance attached to the top end of the verge, which could be used to regulate the early clocks' inconsistent speed. Confronted with this structure, early Japanese clock-makers understood the shape and the function of its various parts by likening them to familiar items or ideas. Thus, for example, they identified a six-tooth gear with another six-sided structure—a snowflake—and dubbed it a "snow(flake) gear."[12] They likened the central role of the crown wheel to the regulatory function of the referee in sumo matches, and named it the "sumo referee gear."[13] The foliot, located at the top of the mechanism and determining its speed, was named "The Seat of Heavens."[14] The index hands were reminiscent of little daggers, and the digits looked like pieces of a *shogi* game.[15] All these gradually became conventional terms, and in the same way that we do not think of actual hands when we talk about the "hands" of the clock, Tokugawa Japanese likely did not actually think of flying daggers when using their term for index hands. In any event, the etymology points to the process of understanding an unfamiliar mechanism by identifying its parts with familiar objects.

Conventions of Dial Layout

By looking at Japanese clock-makers' modifications of European clocks, we also learn that they were inspired by the range of cultural associations various objects evoked. Take, for example, the modification and the placement of hour digits. From our modern perspective, there is a particular spa-

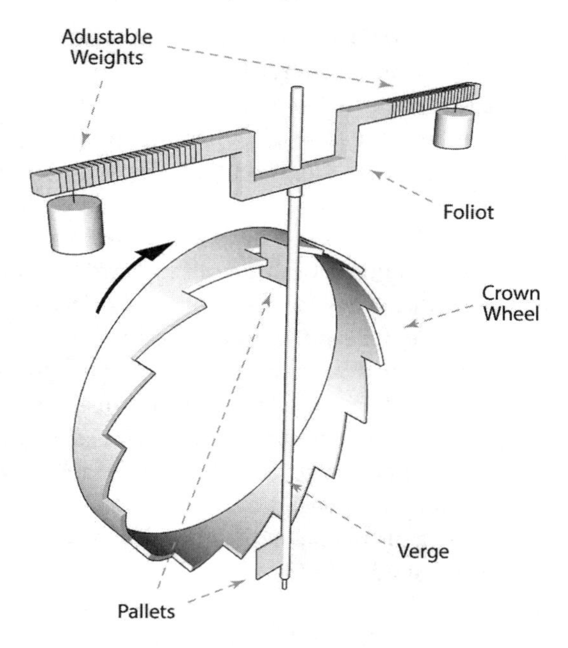

Figure 2.1. Verge-and-foliot escapement mechanism.
The crown wheel of the clock is regulated by the pallets attached to the vertical verge, while the verge is connected to the horizontal foliot. Moving the adjustable weights outwards on the foliot slows the speed at which it rotates, and thus the speed with which the pallets release the teeth of the crown wheel. The crown wheel controls the overall speed of the clock.

tial quality to the arrangement of the digits on the dial. For us, saying that something is "at 2 o'clock" has a very specific meaning. In Tokugawa Japan there was also a spatial association between hour-digits and directions, albeit very different from our own. It was different because it was rooted in a different set of cultural assumptions. In ancient Chinese cosmology the animal signs were identified not only with months and hours, as mentioned earlier, but also with directions. Thus, the sign of the Horse, associated with the summer solstice and with the noon hour, was also associated with south. In a similar manner, the Rat represented winter solstice, midnight, and north. The animal signs were thus also used to indicate directions in maps, which, per local convention, were seldom oriented in such a way that the top of the page signified north. In those maps oriented according to the cardinal directions, north was at the bottom rather than the top. In such maps the Rat sign associated with the north was at the bottom of the page. However, many maps were oriented according to the assumed position of the viewer, so that top of the page was aligned with the viewer's perspective straight ahead, regardless of where the cardinal directions lay. In this kind of map north could be anywhere on the page, depending on the position of the assumed viewer. The assumptions about the placement of the Rat on maps were translated into the placement of the hour-digit of Rat, midnight, on the clock dial. In some clocks the midnight digit was placed where the Rat would be placed on a cardinal-direction map—at the bottom of the dial (see figs. 2.2, 2.9, 2.10). In other clocks, the index hand was fixed in place so as to always be pointing straight up, while the dial with digits arranged counterclockwise on it rotated beneath (see color plates 1, 2, 3). Like the assumed position of the map viewer, who always faces "straight ahead," the index hand indicating "now" always points upward. There were no practical reasons to favor this arrangement of digits over any another. Yet the cultural associations that developed in relation to map-reading practices made the placement of the Rat at the bottom of the dial, or the affixing of the index hand, seem like common sense.

Mechanical Clockwork and Variable Hours

Early Japanese clock-makers' interpretation of the foreign mechanisms was shaped not only by their own cultural associations, but also by their lack of knowledge of European assumptions about clock mechanisms. This is particularly apparent in early adaptations of mechanical clocks to the variable hours system. As mentioned, Japanese clock-makers first encountered clocks that were regulated by a verge-and-foliot escapement mechanism. Early European clocks suffered from numerous mechanical problems that caused variations in the speed of the mechanism and rapidly wore out the gears, permanently affecting the speed. In order to ameliorate these variations in speed, tiny weights were attached to the horizontal foliot (see fig. 2.1). Moving the weights toward the ends of the foliot slowed the rotation of the verge and thus the entire clock; moving them inward toward the middle of the foliot caused the clock to speed up. Consequently, by adjusting the weights, European clock users could maintain a steady speed. Japanese clock-makers, on the other hand, were not aware of the European use of this mechanism. Reverse-engineering the clock, all they could see was that the weights on the foliot caused the mechanism to slow down or speed up, decreasing or increasing the time it took for the index hand to move to the next digit. Thus, viewing the mechanism through the lens of their own timekeeping practices, they realized that the foliot could be used to adjust the speed of the clock in a controlled manner, to make it *purposefully* go

Figure 2.2. A tower clock. The Seiko Museum, Tokyo.
The name of this clock derives from the shape of the pedestal the mechanism was placed on, and which concealed the weights beneath. This particular clock has two foliots, and a dial that places the midnight hour (9, Rat) at the bottom (our 6 o'clock) position.

slower or faster. Namely, the verge-and-foliot mechanism could be used to adjust the clock for seasonal changes. For the long daytime hours of summer, they moved the weights on the foliot outward (slowing down the mechanism), and as the daytime shortened, they gradually moved the weights closer to the verge (to allow the mechanism to run faster).

The Proliferation of Mechanical Clocks during the Tokugawa Period

How did Japanese clock owners know precisely where to position the weights on the foliot? For a long period they did not know, exactly. In the seventeenth century there were no manuals instructing how to move the foliot weights, and there could not have been one, as each clock was idiosyncratic and behaved differently. Instead, a clock-user would gradually learn his (and in the seventeenth century it was mainly "his") clock by tinkering with it and setting it to the ringing of the time-bells in his community. Over time, he developed a feel for a particular timekeeper, gradually getting a sense of where to position the weights on the foliot. This, of course, required constant interaction with the clock. And indeed, in the seventeenth century there were clock-makers themselves who "knew" their clocks, and could properly adjust them. For most of the century, clocks were owned almost exclusively by the wealthy, who could thus afford to hire clock-makers to come and adjust the weights on their foliots twice a day—for daytime and nighttime hours.

This practice changed in the late seventeenth century, the Edo period's cultural "golden era," which saw a sharp increase in the availability of mechanical clocks to a wider public. In 1690, clock-makers and their products appeared in the *Encyclopedia of Professions* (*Jinrin kinmōzui*) in the volume on artisans.[16] Along with knife sharpeners, umbrella makers, and woodblock carvers, clock-makers were just one of the numerous professionals one could see on the streets of Kyoto (see fig. 2.3)

At the same time clocks began appearing in novels. The famous late seventeenth-century writer Ihara Saikaku used clocks in *The Eternal Storehouse of Japan* and *The Life of Amorous Man* as symbols of extravagance and urban pleasures.[17] In a similar manner, the authors of the famous early eighteenth-century puppet play *The Tale of the Forty-Seven Rōnin* used a clock as a rhetorical device.[18] They disguised their pungent critiques of recent political events in the story's putatively medieval setting, yet in the middle of the play they planted a mechanical clock, indicating that the action could only have happened in recent times (see fig. 2.4).

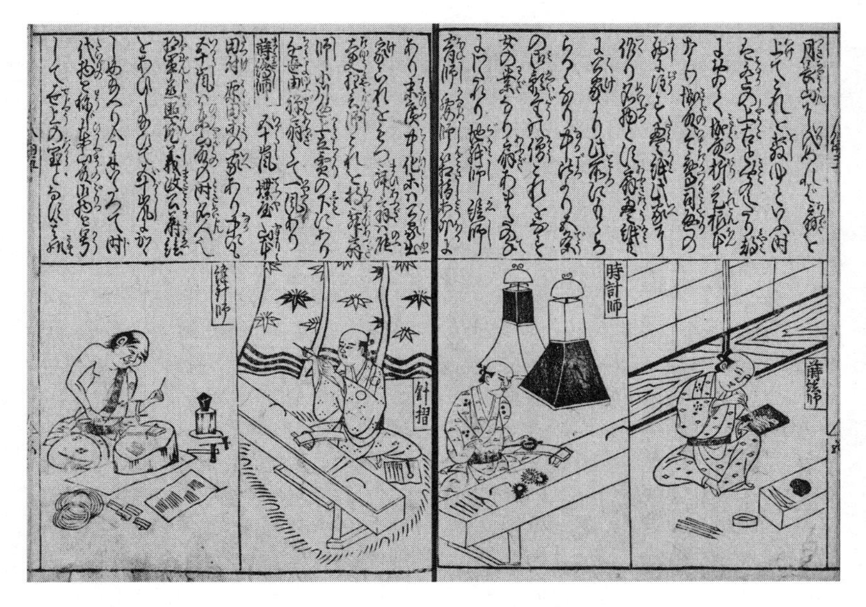

Figure 2.3. "Clock Makers." Fifth scroll from the *Encyclopedia of Professions* (*Jinrin kinmōzui*) published in 1690. National Diet Library Digital Collections. Clock-makers (second from the right) are depicted as skilled in precision work, like lacquer inlay makers (right), needle makers (far left), and needle sharpeners (second from the left). Although appreciated for their rare precision skills, as a profession by the end of the seventeenth century clock-makers themselves were no longer rare.

By the beginning of the eighteenth century, mechanical clocks were no longer rare. Their ownership was no longer restricted to the wealthy few, and although many clocks were luxurious, now there was a growing number of more modest devices. Moreover, thanks to Japanese clock-makers' modifications, they were no longer perceived as foreign. Writing at the beginning of the eighteenth century, the mathematician Nakane Genkei[19] noted that "some clocks imported during the Kanbun era [1661–1672] were locked up in government storage and no one has seen them. But since the final year of the Hōei era [1710], clocks have become extremely popular and widespread."[20] He added: "Even though the clocks were originally brought from the West, nowadays it is impossible to know where they come from, since foreign digits were replaced with our numbers and the animal signs."[21]

As the clocks proliferated, it became more difficult for clock-makers to visit all their clients twice a day for the adjustment of the foliot. Here too clock-makers came up with a solution that showed their detachment from the original assumptions that lay behind the European devices. If there was one foliot that controlled the speed of the clock, could there not be more? After all, the new function of the foliot was to adjust the length of hours measured by the clock according to the seasons, but the length of daytime and the nighttime hours differed. A mechanism with two foliots—one for nighttime hours and another for daytime hours—was a logical result of

Figure 2.4. Illustration from *The Tale of the Forty-Seven Rōnin* (*Kanadehon chūshingura*). The Seiko Museum, Tokyo. By featuring a clock in this scene, the authors signaled that the story actually referred to current affairs, despite its medieval setting.

the process of interpretation and reinterpretations of this European-born technology (see color plate 1). Double-foliot clocks now allowed the adjustment of the foliot weights every mini-season, *sekki*, approximately once every fifteen days.

Towers, Pillows, Steam Buns, and Medicine Boxes

In the early eighteenth century, there were several types of clocks, which the Japanese public identified by their shape. Older clocks, which were set in motion by falling weights, required a large space beneath them. Some of these clocks were attached to vertical surfaces, such as pillars, and were appropriately called "attached clocks"—*kakedokei*[22] (see color plate 1). Others were put on a pedestal inside of which the weights could hang. The pedestals were shaped like castle watchtowers and were dubbed "tower clocks"—*yaguradokei*[23] (see fig. 2.2). During this period, a new type of clock was imported from Holland. This was a box-shaped, spring-driven clock, which lacked weights and hence also lacked the foliot. The square shape reminded Japanese consumers of the square stand used to support one's neck during sleep, and thus these new clocks were named "pillow clocks"—*makuradokei*[24] (see color plate 2).

In addition to these different clocks, the Tokugawa Japanese also enjoyed watches. Watches were imported as early as the middle of the seventeenth century, but like other types of clocks their number increased in the course of the eighteenth century.[25] Early-eighteenth-century sources explicitly state that pillow clocks and watches were all originally imported, and as late as the beginning of the nineteenth century we see statements confirming that "in many cases" the mechanism of "miniature clocks" was foreign and only the dial was changed in Japan.[26] In those cases, apparently, a modified Western mechanism was incorporated into a new case, ornamented with a Japanese-style design, and equipped with a movable-digits dial (discussed below) and a single index hand (see fig. 2.5).

The nicknames given to watches are indicative of the associations they evoked. Alluding to their small round shape, some referred to watches as "steamed-bun clocks."[27] For others, the minuscule yet detailed mechanism was reminiscent of the art of miniature *netsuke* statuettes, and the watches were dubbed "*netsuke* clocks." Because Japanese clothing at the time lacked pockets, small items were carried in one's long sleeves, and watches also became known as "sleeve clocks"—*sodedokei*.[28] Or a watch could be incorporated into a portable medicine box (usually carried on one's belt) and was thus called "a medicine box clock"—*inrōdokei*[29] (see color plate 3).

Figure 2.5. A Western watch mechanism with a modified Japanese dial. The Seiko Museum, Tokyo.
The inner mechanism of this watch was made in London and imported to Japan. Japanese clockmakers removed the minute index hand, and adjusted the gears so that the hour hand only made one revolution per day. Then, they removed the Western dial and affixed a Japanese dial with movable digits.

Pillow clocks and watches forced Japanese clock-makers to come up with a new method for measuring variable hours. Both were spring-driven and had "deadbeat" escapements that lacked foliots, meaning there was no place to attach weights to change the speed of the mechanism. In order to make European devices usable, a few Japanese clock-makers artificially attached a foliot or two to their spring mechanisms to allow for a familiar mode of adjustment. More, however, found another way to adapt the new technology to Japanese conventions of time measurement. Instead of forcing the index hand of the clock to move faster or slower as it traveled between the digits, clock-makers made the digits themselves move by adjusting the amount of space between them. The small movable plates on which the hour digits were placed were visually reminiscent of the pieces on a *shogi* board, and since they were now split, they were dubbed "divided *shogi* pieces," *warigoma*[30] (see color plates 2, 3, 10).

Although it was practical necessity that forced Japanese clock-makers to come up with a new solution for measuring variable hours, the *kinds* of solutions they devised were shaped by their culture-specific associations. In particular, they found inspiration in nonmechanical timepieces, such as the incense clock described in the previous chapter. In the incense clock, the incense burned at a steady rate, yet the signposts indicating the hours could be placed at different intervals, reflecting how the length of hours varied with the seasons.[31] Being familiar with the way incense clocks measured hours of changing length, Japanese clock-makers sought to employ a similar strategy in mechanical clocks. They were used to visualizing hour-digits as movable, and saw no problem making the digits of the mechanical clock faces movable as well. Not only were they not restricted by the association of digits with a particular place on the dial, but they also possessed examples that demonstrated to them how digits could, in fact, change position throughout the year.

Graphs and Dials

Another example of design-by-association can be seen in a vertical clock invented in Japan and named a *shaku* clock[32] (see figs. 2.6, 2.7). A *shaku* was a unit of length measuring about one foot, but the *shaku* clocks were not named for their size, which varied, ranging from several feet to a couple of inches. Rather their name derived from their resemblance to a measuring tool, a ruler, used to measure a length of one *shaku*. This ruler was also visually similar to the notched arrow or rod held by the figurine in the ancient clepsydra described in the previous chapter. The *shaku* clock looked very like the clepsydra rod and required a similar way of viewing. It was vertical, it was marked with notches, and the index hand moved daily from the top to the bottom. The earliest clocks of this kind were rather simple—a falling weight to drive the mechanism was put in front of a measuring rod and thus double-functioned as an index hand. When the movable-digits solution became popular, *shaku* clocks, too, were made in such a way as to allow the digits to move up and down (see fig. 2.6). Yet another type of *shaku* clock simply had replacement plates for every season[33] (see fig. 2.7). Associations with familiar practices—and familiar timepieces—thus inspired new clock designs. Japanese clock-makers were guided by not only the question "How do I make a clock that measures variable hours?"

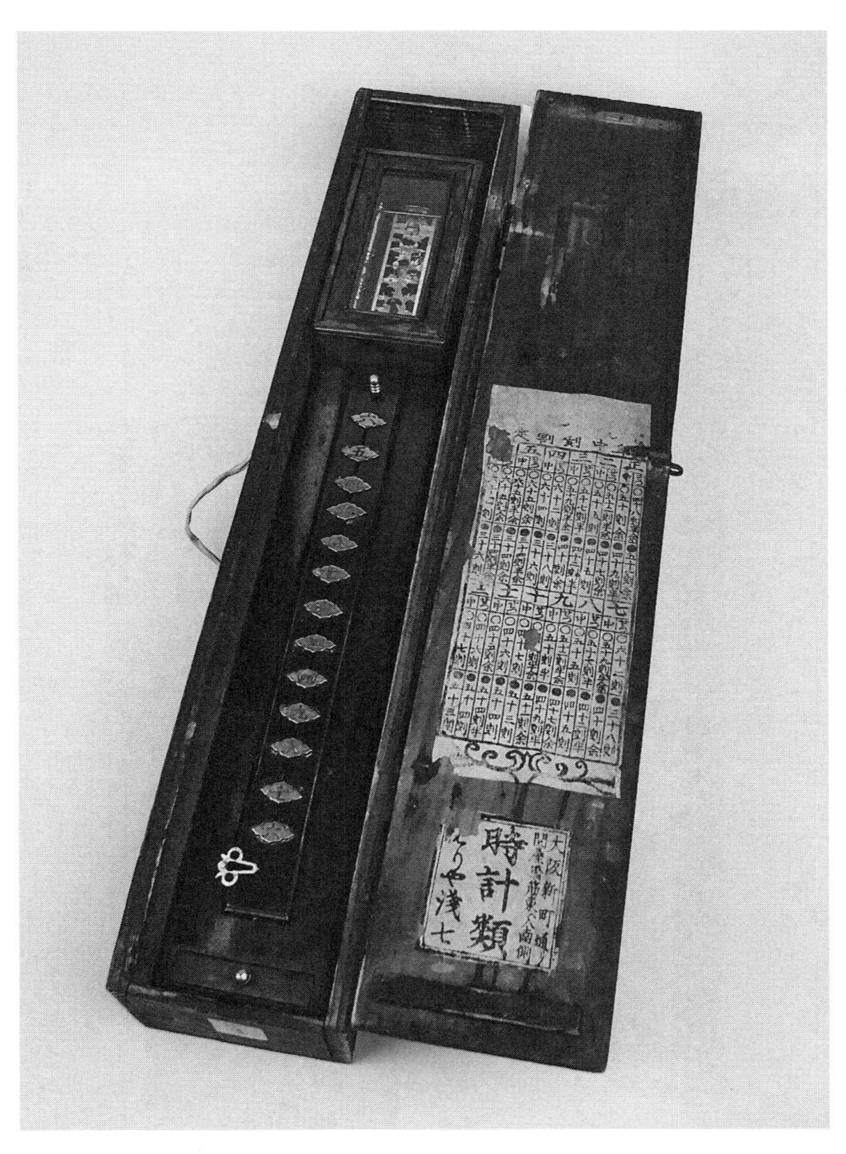

Figure 2.6. A *shaku* clock with movable digits and an attached time-table (right, inside lid). The Seiko Museum, Tokyo. The owner of this clock would consult the table in order to manually adjust the movable digits according to the season. The table was divided into twelve months, which were further divided into half-months, forming twenty-four possible variations. For each season, the table showed how many of the hundred equal *koku* units constituted a nighttime period and how many a daytime period. It was up to the owner to imagine the hundred-*koku* division on the vertical dial, and to adjust the movable digits according to the table.

Figure 2.7. A *shaku* clock with four replacement plates. The Seiko Museum, Tokyo. The index hand moves vertically along the dial at the same rate. However, instead of adjusting the movable digits, the owner of this clock would replace the whole dial according to the season. Placed side by side, the replacement plates form a graph.

but also "How do I make a mechanical device that measures the hours in the same manner as the incense clock or the clepsydra?"

In designing their mechanical clocks, Tokugawa-period clock-makers relied not only on their knowledge of familiar nonmechanical timepieces, but also of timekeeping accessories such as seasonal charts and graphs. The basic annual calendar was already a kind of chart (see fig. 1.4). Seasonal hour charts relied on the format of the calendar and encoded the length of the variable hours in each of the twenty-four seasons, placing hours on what we would call the y-axis, and seasons on the x-axis. While detailed charts were used mainly by bell keepers, simplified charts were more common and printed within many travel guides. Seasonal charts, in particular, were often used to create paper sundials. These sundials consisted of a set of paper gnomons of different lengths for different seasons (the x-axis), and a base with lines indicating hours (the y-axis). In order to know the time, the user would face north, lift the paper gnomon for the given season, and estimate the length of the shadow relatively to the markings on the base of the sundial. Together, these markings formed a graph (see fig. 2.8).

The charts and the graphs were probably the source of inspiration behind the graphlike *shaku* clock (see color plate 4). The dial of this clock had columns for the different seasons, with hour markings arranged at appropriate distances. It required the user to trace two kinds of movement: the daily top-to-bottom movement of the horizontal plank, and the seasonal movement of the bob placed on the plank from right to left and vice versa. The owner of the clock manually adjusted the location of the bob according to the seasons inscribed at the top of the graph and then located the position of the bob relative to the graph inscribed on the dial that showed the change in the length of hours over the year. This largely replicated both the appearance and the use of the paper sundial, though the *shaku* clocks no longer required any analog sensitivity on the part of the user. As long as one knew what time of the year it was, the design of the dial would show the correct seasonal time on its own.

A fully automated clock design too was based on a graph. In the very last years of the Tokugawa period, Japanese clock-makers came up with round graphlike clocks, which automatically switched between seasonal times (see color plate 5). Instead of columns for the different seasons, these clocks had concentric circles, and instead of a horizontal rod with a bulb that served as an index, they had an index hand that changed its length with the seasons. Attached to a spiral groove, the index hand could move closer or farther from the center. In the winter, it was barely visible, with only its tip peeking out from underneath the central plate. In the summer,

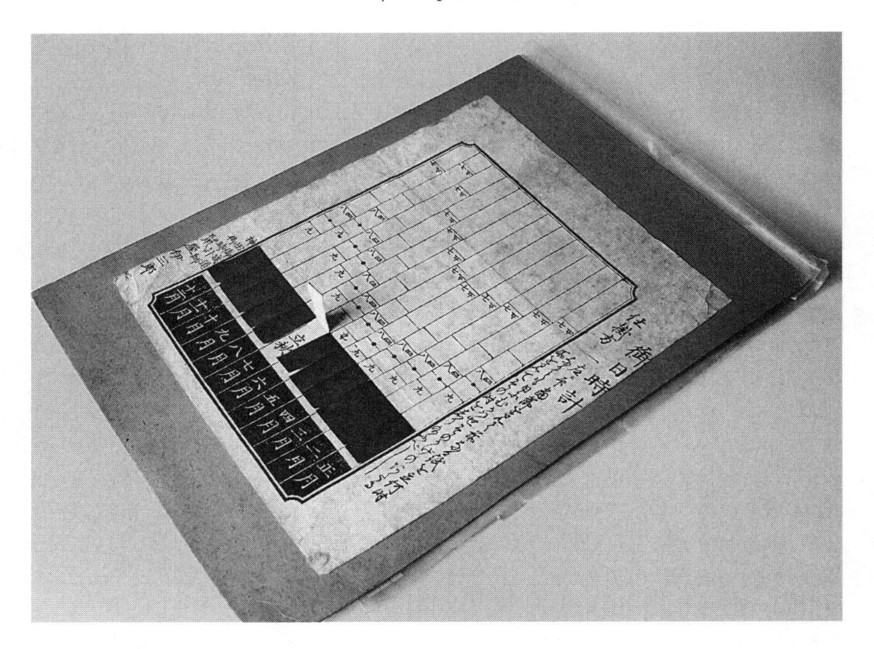

Figure 2.8. A paper sundial with a seasonal hour diagram. The Seiko Museum, Tokyo. Twelve strips of paper correspond to the twelve months and twelve out of the twenty-four seasons. Each strip can serve as a gnomon. Facing north, the user would fold the strip of paper appropriate to the season into gnomon position and identify the length of the shadow relatively to the chart printed on the paper below. The chart indicates the length of the shadow at noon (9th hour), one Tokugawa hour before and after the noon (4th and 8th hours), and two Tokugawa hours before and after noon (5th and 7th hours)—thus covering all the daytime hours in the variable hour system. Side by side, the seasonal markings of the hours form a graph with curving lines. Depicted in this image is a gnomon lifted to the season called "Beginning of Autumn," in the seventh month according to the lunar year calendar (approximately August). The shadow stretches just beyond the 4th or the 8th hour.

the index hand extended all the way to the edge of the dial. In order to ascertain the time, the user would look at the precise location of the tip of the index and note its position relative to the hour-lines, as if the clock dial were combined of several seasonal dials arranged as concentric circles. On the back of one of these clocks, made in 1834 (Tenpō 5), the clock-maker inscribed the following message:

> [Clocks] are so widespread, they are used everywhere. Nowadays, there is no professional who doesn't rely on them. However, the hours of the different seasons are not equal. Consequently, you need to add and subtract every day, based on the hours of the Rabbit [dawn] and the Rooster [dusk]. But determining the dawn and the dusk is difficult and everybody is troubled. . . .

> By calculating the times of the *sekki* and the *kō*, aligning those times with ones displayed by the clock, and making them move, [I] created a clock that shows the seasonal time on its own, matching the [time of the] heavens.[34]

Here too, the automation of the clock was a necessity, yet the solutions emerged out of clock-makers' existing associations with other timekeeping practices. In designing this clock, the clock-maker obviously relied on the fact that the users would already have had experience with graphlike clocks, and would know how to spot a cross-section between the index-hand and the relevant column. The exact visual appearance of the dial was inspired by another familiar object—the graph made to visually represent the yearly changes in the length of daytime and nighttime hours (see fig. 2.9). In this visual representation, concentric circles represented seasons. All the circles were equally divided into one hundred astronomical *koku*, with some *koku* colored white to indicate daytime and others colored black to indicate night. Round marks on certain *koku* indicated the beginning of an hour during each season. Combined together, these marks formed curved lines. These same lines were reproduced on the face of the round graphlike clock. And in order to read the time off this clock, the user would have to employ the same method as when looking at the paper graph—identify the "now" in between circles and the curved lines. The automation was a mechanical solution, but it was made to accommodate a design inspired by an existing visual image of seasonal changes in the amount of daylight and an existing practice of tracing the "now" on such image.

Mechanical clocks never replaced nonmechanical timepieces during the Tokugawa period. As mechanical clocks proliferated and became more widespread and affordable, so did nonmechanical timepieces. In the beginning of the nineteenth century, many Tokugawa Japanese owned and carried around timepieces—some mechanical and others not. When Philipp Franz von Siebold stayed in Japan in 1820s, he recorded the variety of Japanese timepieces he encountered or saw in books (see fig. 2.10). The impression he had was that there were numerous timekeeping technologies in Japan, each with its own idiosyncratic way of displaying the passage of time.

* * *

These various timekeeping technologies did not just coexist—they informed the conceptual world of Tokugawa-period clock-users and makers. These various devices provided tangible material examples of such abstract notions as the movement of time, or its spatial mapping. They informed Tokugawa users' assumptions about how to handle timepieces, where to look in order

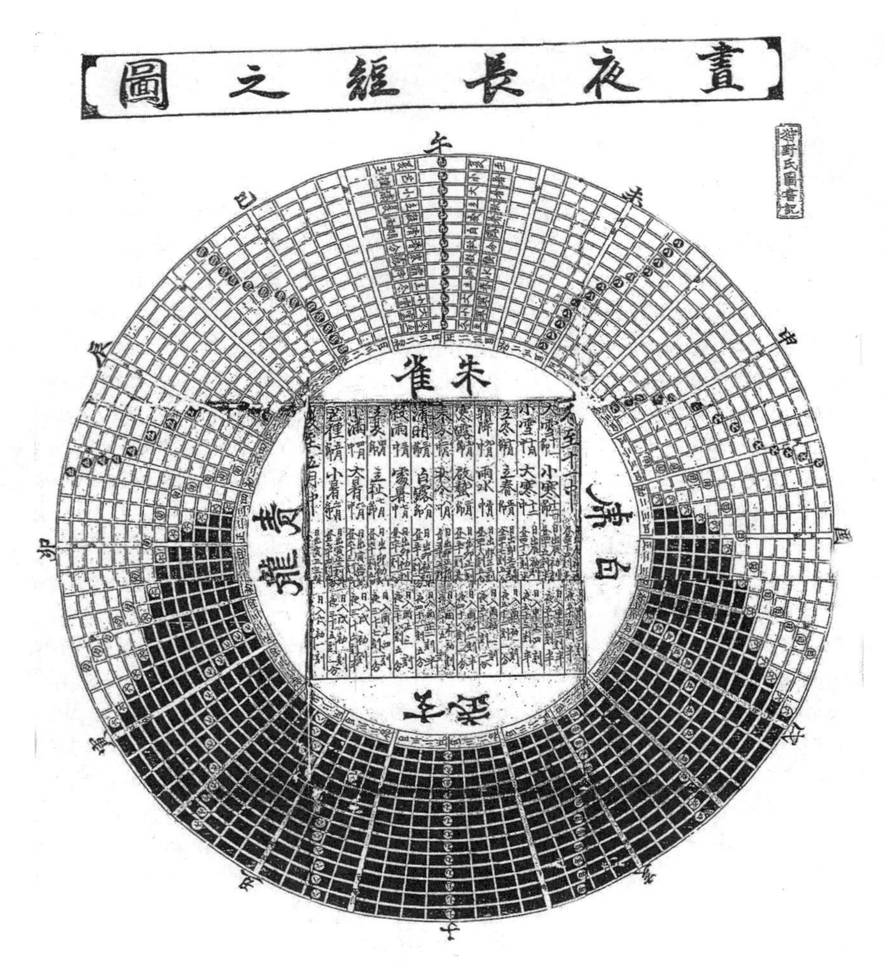

Figure 2.9. *Illustration of the Length of Days and Nights* (1720).
Tōhoku University Kanō Archive, Sendai.
In this circular graph each concentric circle represents a pair of seasons (from winter solstice to summer solstice and back). Each circle is divided into one hundred segments that correlate to the hundred astronomical *koku*. The white segments represent daytime, the black ones represent nighttime, and the tiny circles within some of the segments represent the beginning of an hour. In order to determine when during a variable-hour day a certain astronomical *koku* occurs, one would identify the segment corresponding to the given *koku* and the circle corresponding to the season. Thus, for example, the same *koku* corresponds to the 7th hour of night (Tiger) during the winter, the 6th dawn hour (Rabbit) during the spring, and the 5th daytime hour (Dragon) during the summer. Working with this graph, the reader would develop the habit of identifying a particular segment, rather than imagining hours as uniformly shaped regions defined by radial boundaries.

Figure 2.10. Philipp Franz von Siebold, *Nippon, Archiv zur Beschreibung von Japan*, "Timepieces."
Siebold's depiction of timepieces available in Japan shows a tower clock (Fig. 7); a *shaku* clock (Fig. 9); a clock dial, and a visual representation of hours with the midnight hour (Rat) at the bottom (Figs. 8 and 10); a paper sundial (Fig. 11); clepsydras (Figs. 1, 2, 3); as well as other devices. Judging from differences in the style of representation, it seems that the *shaku* clock, the round clock dial, and the paper sundial were the only actual objects Siebold drew. The others are likely copies of illustrations from Japanese books, such as *Illustrations of Various Measurement Devices* by Nishimura Tōsato.

to locate the "now," and where one could expect to find hour-digits. They formed an array of associations related to the practice of timekeeping.

Associations with familiar objects provided Japanese clock-makers with conceptual tools to approach mechanical timekeeping technology. From the very beginning, clock-makers comprehended the mechanism by likening its function and appearance to objects familiar to them from their own daily lives. Once they identified European devices as belonging to the category of timepieces—*tokei*—they interpreted the dial and the mechanism through the lens of what they knew about other kinds of *tokei*. Thus, when they modified the mechanism, adjusted the dial, or invented entirely new mechanical timepieces, they did so on the basis of their existing associations with timekeeping terms and functions.

Looking at clock-makers' craft from the standpoint of their associations, we understand not only why they sought *a* solution to one or another practical problem, but also *how* they came up with the *kinds* of solutions they did. The need to find a solution may have been posed by incompatibility between the structure of a device and existing timekeeping practices, but the solution itself was neither technologically nor sociologically predetermined. Numerous strategies to make mechanical clocks measure variable hours were motivated by clock-makers' familiarity with methods of telling the time using nonmechanical timepieces, or using graphs and charts that visually represented seasonal variations. Japanese clock-makers not only integrated mechanical clocks into Tokugawa social structures by making them measure variable hours—they also integrated the clock into the world of their particular cultural associations by making clocks measure time in a manner similar to that of the incense clock, the clepsydra, and the paper sundial.

Such processes of interpretation and modification of technology change our perspective on the phenomenon commonly referred to as "technology and knowledge transfer." It is true that a given physical object left the coast of Portugal or Holland and arrived in Japan. Yet what Japanese clock-makers and clock-users saw in the material structure of this device depended on their own cultural background, on the variety of objects they knew, and on the ways these objects were used. Seeing the original interpretation of mechanical clocks in Japan, can we confidently say that it was the "same" technology as in Europe? Or was it something more akin to a resource, which could be molded into a technology through a process of interpretation? As we shall see in the next chapter, for the same technology to be "transferred," it needed to come with a related body of knowledge, and furthermore, with a related array of associations.

THREE

Astronomical Time Measurement and Changing Conceptions of Time

Time Keepers and Timekeepers

Although timekeeping practices were changing across different social groups, it was the transformations in astronomers' practices that inspired an appreciation of Western-style timekeeping and eventually led to the epochal 1873 calendrical reform. Astronomers were by no means representative of the typical Tokugawa-period clock user. Astronomers' timekeeping practices—the reasons they measured time, their temporal units, and their timepieces—differed significantly from those employed in everyday life. Most Japanese were probably not even aware of how astronomers measured time. Yet despite being far from representative of Tokugawa society, astronomers nevertheless played a crucial role in changing the public perception of time measurement—not only because they determined the variable hours on which everybody relied, but also because they influenced would-be reformers who demanded the adoption of Western timekeeping in the middle of the nineteenth century.

An astronomer's job was to study the heavens for the purpose of improving mathematical algorithms for devising the official calendar. The goal of finding the perfect calendrical algorithm was an unattainable task. The parameters that influence the motion of celestial bodies are simply too numerous, and no matter how precise and complex the algorithm is, over time it will inevitably diverge from observed reality. Today, we have given up the idea that a calendar needs to account for all the plethora of celestial motions, and we simply add leap seconds here and there to compensate for changes that creep in. Yet multiple generations of Japanese astronomers aspired to find a perfect astronomical algorithm, only to discover later that what they thought of as "perfect" turned out to be very slightly imprecise. To quote an analogy offered by the famous Japanese historian of astronomy Shigeru Nakayama, "If somebody drives a totally straight highway with even the tiniest inclination, they will drive off the road, eventually."[1]

In Japan as elsewhere, astronomers' work sometimes manifested in calendrical reforms, which were customarily justified by claims that the previous calendar had gone out of sync with natural events. Yet there are also numerous documented instances of calendars that were allowed to remain asynchronous with observed celestial events. In Japan, the calendar was not reformed for eight hundred years prior to 1685, and in Europe it took even longer to reform the Julian calendar, despite its shortcomings. On the other hand, during various eras in China, calendrical reforms were very frequent, sometimes introducing only superficial changes to algorithms that generally worked pretty well. Decisions to reform the calendar were always political, yet the changes they introduced often reflected ongoing astronomical work that happened outside of the political realm.

Astronomers' views of time measurement were shaped by the ways they employed measured time. In their practices, astronomers needed two basic kinds of data—the location of a given phenomenon (such as an eclipse), and the precise moment of its occurrence. Consequently, many of their tasks focused on two types of measurement that often had to be done simultaneously—a measurement of space and a measurement of time. Today, we generally think of the latter in quantitative terms—the more precise the better. Perhaps we assume that time measurement is exclusively quantitative because we have only one timekeeping system for all of our purposes. The evolution of Tokugawa-period astronomical timepieces and timekeeping systems, on the other hand, reveals the fact that time measurement is never neutral, but rather is shaped by the particular details of the intended use of the time being measured. The mathematical procedures for which astronomers needed to measure time, the visual representations of measured values, and the material environments surrounding acts of measurement all shaped the ways astronomers perceived their practices of measuring time. As we shall see in the following pages, changes in astronomical practices also brought about a transformation in astronomers' perception of time measurement. As they incorporated more elements of European astronomy into their work, they began treating time measurement in a way that was increasingly similar to that of their European colleagues.

Local Time and the Jōkyō Reform

As noted, in the eight centuries preceding 1685, no changes were made to Japan's calendrical algorithm. The calendrical system had been adopted from China in the seventh century, together with other forms of governance. The overall system was modified to incorporate local practices and beliefs, yet the calendrical algorithm itself was adopted as is. Trying several different Chinese algorithms, Japanese rulers finally settled on the one employed in the Senmyō calendar of AD 862.[2] From that year on, the same calendrical algorithm was used until the mid-seventeenth century, when Hoshina Masayuki instigated a reform.[3] Masayuki was a half-brother of the third Tokugawa shogun, Iemitsu.[4] When the latter died in 1651, Masayuki was appointed to serve as a regent to his ten-year-old nephew, child-shogun Ietsuna.[5] Masayuki learned that throughout history Chinese rulers had used reforms—and particularly calendrical reforms—to solidify their reign. He hoped that by reforming the calendar and imposing it on the provincial domains, he would make them dependent on the central government and thereby strengthen Ietsuna's rule.

Masayuki chose the relatively unknown Yasui Santetsu the second to carry out the reform. Yasui Santetsu the first—the latter's father—had been a professional *go* player hired to play the strategic game with the shogunal family. Santetsu the second was educated alongside the children of the broader shogunal family and upon his father's death assumed the role of shogunal *go* player. *Go* requires exceptional skill in calculating probabilities—an asset possessed by both father and son (the pseudonym Santetsu translates literally as "mathematical wisdom"). In the eyes of Masayuki, the combination of mathematical excellence and an intimate relationship with the shogunal family made Santetsu the second an ideal candidate—both skilled and loyal. As a reward for his service, the former *go* player was bestowed the name by which he is now commonly known—Shibukawa Shunkai (or Harumi).[6]

Shunkai based his calendrical algorithm on a thirteenth-century Chinese calendar titled *Granting the Seasons* (*Shoushili*, or *Jujireki* in Japanese).[7] He first learned about this centuries-old system from a Korean scholar, Yō Razan, and was immediately impressed by its elaborate mathematical methods.[8] For a former *go* player who had been trained to calculate probabilities, mathematics was the primary factor in his choice of a model astronomical system, and *Shoushili* offered the most methods and the most numerical data of any system known in Japan at the time.

The other possible astronomical models that Shunkai had encountered had not met his standards. He studied the more recent calendrical algorithms from Ming China, but concluded that they offered only cosmetic changes to *Shoushili*, or simplified the elaborate methods he valued so highly. He was also aware of Western astronomy, but the little knowledge that had trickled into Japan was too meager and too superficial to serve as a basis for calendrical reform. There were several texts written by former Japanese converts who were forced to renounce Christianity and practiced astronomy as a secret form of devotion.[9] The former Jesuit Chiristóvão Ferreira (aka Sawano Chūan), authored a spoken-Japanese explanation of Ptolemaic astronomy that was transcribed and translated into classical Chinese by a local scholar.[10] There was also the popular *Questions about Heavens* (*Tianjing huowen*), written by a non-astronomer student of Jesuits in China.[11] Jesuit literature from China was officially banned in Japan throughout the seventeenth century. Strictly speaking, even mentioning the name of the most famous Jesuit active in China, Matteo Ricci, would have resulted in having one's copy of his book confiscated by Japanese authorities. In practice, however, many such books—including *Questions about Heavens*—were smuggled into Japan and openly referenced in local literature.[12] The common characteristic of all the above texts on Western astronomy was their exclusive focus on cosmology. They ignited the imagination and offered visual representations that sometimes found their way into the popular Japanese literature of the early eighteenth century. But they ignored observational and calculational procedures and provided little or no data that could be integrated in the practices of professional Japanese astronomers. Shibukawa Shunkai's writings reveal that he was well acquainted with this literature.[13] Yet without numerical data and information on actual astronomical methods, he concluded that "even though I understand the logic of Westerners, it is impossible to put it in to practice."[14]

In contrast, *Shoushili*, or *Granting the Seasons*, provided a wealth of numerical data from past observations, a detailed explanation of computational methods, various observational procedures, and descriptions of a variety of observational instruments that could be used to generate more data. For example, it offered much more precise values for astronomical constants, such as the difference between the lengths of a tropical year (365.2425 days) and a sidereal year (365.2575 days).[15] It described the regularity of variations in the motions of the sun, the moon, and the planets, as well as where celestial bodies were expected to accelerate and decelerate in their orbits.[16] It discussed whether calculations should be made according to the apparent or the mean motions of the sun and the moon—a crucial question given that most astronomical phenomena do not fall neatly into definition by familiar units such as day, month, or year.[17] It suggested a new way of determining the exact time of the winter solstice. Above all, it offered new mathematical methods that relied not only on the computation of intervals between occurrences of the same celestial phenomena but also the geometric relationships of spherical planes, introducing the notion of the mathematical arc.[18]

The authors of the *Granting the Seasons* system devised numerous instruments to generate data that would be used in these calculations.[19] Eleven of seventeen instruments in the observatory made for the *Granting the Seasons* reform were related to time measurement, supplementing extant instruments such as the clepsydras located in the inner chambers.[20] Timekeeping instruments were specifically designed for a variety of astronomical practices related to time measurement.[21] Clepsydras were intended to measure time during nocturnal observations, dripping at a constant speed. Enormous gnomons—one eight feet long, another forty feet long—were used to determine the length of the sun's shadows used in the calculation of the precise moments of the solstices. An "upward-facing instrument" was a semi-spherical sundial that had multiple dials for the different seasons.[22] The pinhole in its gnomon served both to narrow the shadow and to reflect the stages of a solar eclipse, enabling both the observation and the timing of an eclipse in one single action. Finally, the "simplified instrument" was a disintegrated armillary sphere—a device with a series of non-concentric rings for determining such parameters as the declination of stars, celestial equators, and celestial poles[23] (see fig. 3.1). One of those rings was an equatorial sundial called a "hundred-*ke* ring"—referring to the hundred daily "notches of the clepsydra" called *koku* in Japanese and *ke* in the Chinese original.[24] The sundial ring was stable on the instrument but was tilted to reflect the latitude of Beijing, with the gnomon pointing to the celestial pole. The upper part of the ring had markings that told the hours between the spring and the autumn equinoxes while the bottom part told the time during the second half of the year.

Admiring the *Granting the Seasons* system, Shunkai adopted the majority of its methods, but also introduced conceptually significant modifications rooted in his particular worldview. In addition to studying the skills necessary for astronomical observation, he studied philosophy and cosmology under Yamazaki Ansai.[25] Ansai developed a peculiar philosophy that combined the neo-Confucian metaphysics of Zhu Xi (via the teachings of Korean philosopher Yi T'oegye [1507–1570]) with Japanese traditions and history. This combination developed into a particular brand of Shinto—*Suika Shinto*—that emphasized the structural harmony of the apparently

Figure 3.1. The "simplified instrument." Ming-period replica. Beijing Astrological Museum. Photograph courtesy of Professor April Hughes. Unlike the armillary sphere, the "simplified instrument" is not meant to serve as a model of the universe nor is it a predictive device. Rather, it is an assemblage of observational tools. The hundred-*ke* ring sundial is on the right in this image, tilted to reflect the specific latitude of the observatory from which *Granting the Seasons* originates.

diverse universe, and that drew connections between the workings of the heavens and the deeds of individual humans. This cosmological quest for harmony between the universal and the particular motivated Shunkai's modifications of the Chinese *Granting the Seasons* system.[26]

The most prominent feature of Shunkai's calendrical system was its emphasis on the geographical characteristics of the Japanese archipelago. Following Yamazaki Ansai and other scholars of the late seventeenth century, Shunkai demanded that Chinese systems be modified to accommodate the reality of Yamato (an ancient name for Japan), and even named his own system the "Yamato calendar" (*Yamato reki*).[27] One crucial difference between China and Yamato was in the timing of the seasons. Shunkai was one of the first to openly discuss the fact that many of the Chinese names for the seasons (*sekki* and *kō*, see chapter 1) did not correspond to actual seasonal changes in Japan. He recommended observing the occurrences of natural phenomena in Japan and revising the system accordingly.[28]

Another major problem with implementing the *Granting the Seasons system* in Japan was the fact that its calculations were based on the location of the observatory in Beijing—far to the west and almost five degrees north of the latitude of Kyoto. The treatise casually mentioned the notion of "difference in leagues," *risa*, which indicated that observations in different locations would produce different results.[29] Shunkai picked up on this notion, stressing that "difference in leagues" translated into differences in time measurement. The difference in longitude between China and Japan resulted in what Shunkai called a "difference in the point of observation" (*shisa*).[30] To use the numerical data from *Granting the Seasons* in Japan, one had to correct for longitudinal difference. Shunkai estimated that a distance of 150 *ri* resulted in one *koku* (one-hundredth of a day) difference in time, and concluded that even within Japan itself there could be observational differences of up to five *koku*. Latitude presented another problem. The differences in solar angle at various latitudes resulted in gnomon shadows of different lengths and also in different perceived durations of day and night.

The solution Shunkai found was to adjust the way he measured time. He modeled his timepiece on the hundred-*ke* ring—in Japanese, the hundred-*koku* ring (see fig. 3.2). However, noting that the instrument described in *Granting the Seasons* was stable at a degree corresponding to Beijing's latitude, Shunkai made his *Ring* movable, and hence adjustable to any degree of latitude. Describing this instrument, Shunkai's student Tani Shinzan noted that the handles on the sides of the device enabled him to tilt the ring "according to every country," rendering the device usable in "all the myriad countries."[31]

The time measurement allowed by this particular timepiece reflected Shunkai's metaphysical convictions. The hundred-*koku* ring allowed him to measure time at the latitude appropriate to Japan, as distinct from China, emphasizing the centrality of Yamato. It also conveyed the sense that time was both universal (the device could be used anywhere) and bound by locality (it had to be adjusted to the particular latitude of a place). This notion reflected the harmonic relationship between the universal and particular central to Suika Shinto, taught by Yamazaki Ansai to young Shunkai.

Unfortunately, Shunkai's cosmological views also had a negative impact on his mathematics. His enthusiasm for harmony made him averse to unattractive fractions. Thus, for example, he emphasized the so-called Law of Vicissitudes, according to which the length of the year was slowly but constantly diminishing.[32] *Granting the Seasons* established that every one hundred years, the length of the year decreased by 2/10,000. Shunkai subsequently deduced that every year is shorter by 0.000002 of a year, or by 0.000730485 of a day.[33] This, evidently, seemed to Shunkai to be a too

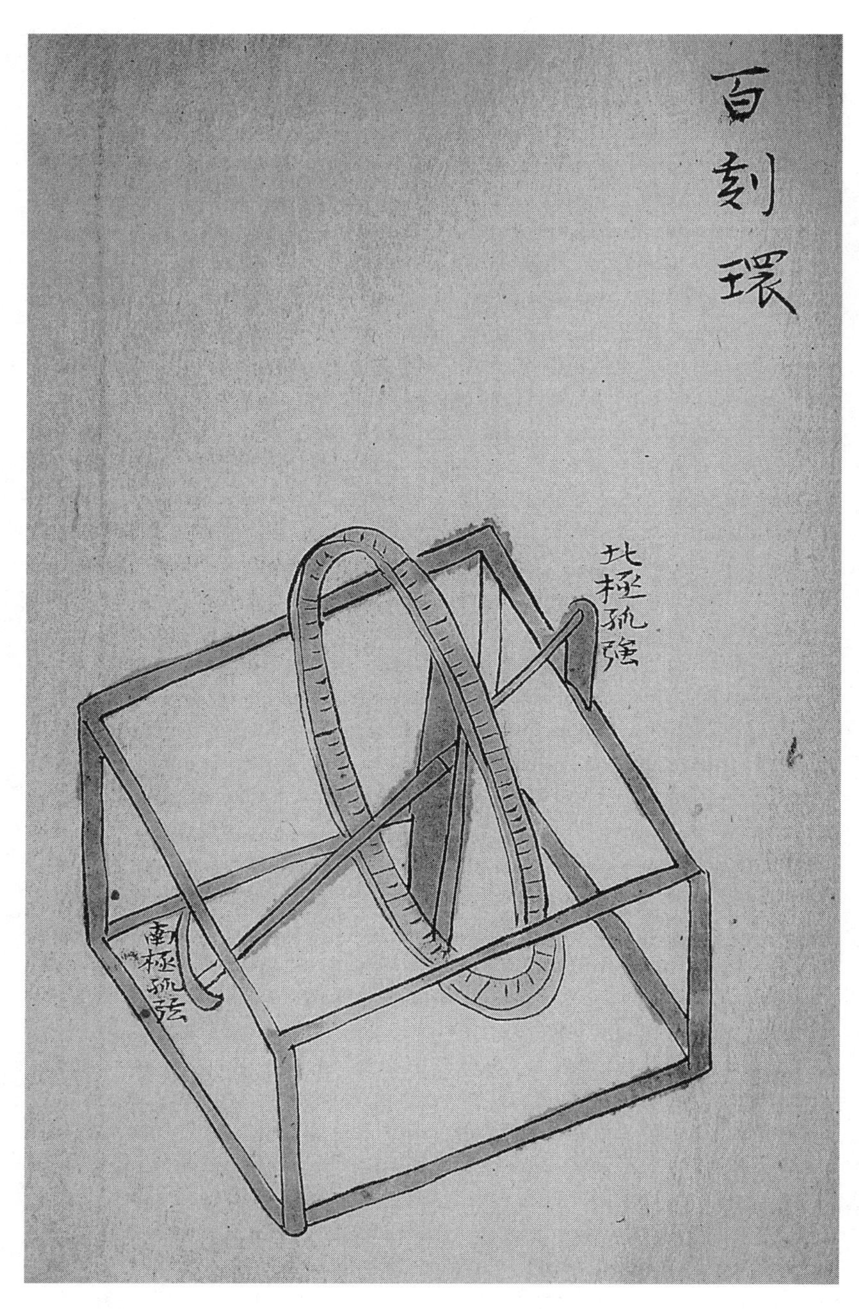

Figure 3.2. Nishimura Tōsato's sketch of Shibukawa Shunkai's hundred-*koku* ring timepiece. *Illustrations of Various Measurement Devices* (*Sokuryō shoki zu*), date unknown, second half of the eighteenth century. National Astronomical Observatory of Japan, Mitaka. Tōsato unsuccessfully attempted to reconstruct Shunkai's device and eventually gave up on the instrument, writing that he "would be only happy if somebody later would understand what Shunkai said and prove me wrong."

crude and seemingly nonharmonious value, and he decided to round it up to the exact measure of one ten-thousandth of a day—an abstract unit, *byō*, which was one-tenth of a *bu*, which in turn was one-tenth of a *koku*.[34]

It took several decades for Shunkai's calendrical system to finally be implemented. In part this was due to the fact that his rounding up of numbers resulted in embarrassing failures, as when his system proved worse at predicting eclipses than the eight-hundred-year-old algorithm it was supposed to replace. More devastating to him, however, were the deaths of Hoshina Masayuki, in 1673, and then, in 1680, of the shogun Tokugawa Ietsuna, and the political intrigue that followed.

Despite these obstacles, the reform was carried out in 1685, the second year of the Jōkyō era, and had wide-ranging ramifications.[35] The reform centralized calendar production, forcing the domains to depend on the shogunate in all matters of time management. It thus fulfilled the goal originally envisioned by Hoshina Masayuki, albeit benefiting the next shogun, Tsunayoshi, who ruled after Ietsuna's death.

Shunkai recalculated the whole system to adjust it to the geographic location of Kyoto. In doing so, he was partially motivated by practical considerations—the main observatory was located near Kyoto. But he was also driven by ideological and metaphysical convictions. His teacher Yamazaki Ansai had emphasized the centrality of Japanese cultural history, which focused on the ancient capital of Kyoto (formerly Heian). Even though the capital moved to Edo with the establishment of the Tokugawa shogunate, Kyoto was still considered the cultural heart of Japan. By basing the calendar on local time there, the reform reinforced the idea that the whole country was run on the same time.[36]

Mathematics and the Importation of Western Astronomical Literature

One of the unintended outcomes of Shibukawa Shunkai's work was an increased interest in Western astronomy. Interestingly enough, it was his engagement with the thirteenth-century Chinese calendar that paved the way for the acceptance of Western astronomical practices in Japan. His reliance on the *Granting the Seasons* system in his calendrical reform brought this ancient treatise to the attention of Japanese mathematicians, such as the famous Seki Takakazu, as well as Takebe Katahiro and Nakane Genkei, who are less known today but were quite prominent during the period.[37] Takebe, in particular, reconstructed the mathematical methods of *Granting the Seasons*, which were described in somewhat obscure language in the original thirteenth-century text.[38]

The feature that most intrigued mathematicians like Takebe was the system's treatment of triangles. Though the mathematical calculations were not exactly "trigonometry" as we know it today, they did involve calculating the relationships between angles and the lengths of sides of triangles. More importantly, *Granting the Seasons*' mathematics discussed problems in which one side of the triangle was an arc. These early forms of trigonometry inspired Japanese mathematicians to attempt to further develop Chinese methods.

Takebe and others quickly recognized that Western astronomical treatises relied on similar methods and thus could provide answers to mathematical conundrums. But access to Western astronomical writings at the time was extremely limited. There were just a couple of works on Western astronomy available in Japan—popular texts that abbreviated detailed explanations of mathematical methods and rounded up (or down) multidigit numerical values. One of these works, *Questions about Heavens* (*Tainjing huowen*), referenced other Chinese treatises on Western astronomy's mathematical methods, such as *Calendrical Book Expounding New Western Methods* (*Xiang xinfa lishu*).[39] Such works were difficult to get because they either were written by or mentioned Jesuits, and thus were forbidden as "Christian propaganda." Increased access to the desired literature on European mathematical methods would require a policy change.

Access to literature on European astronomy became possible when the eighth shogun, Tokugawa Yoshimune, assumed power in 1716. By that time, Yoshimune had encountered Western astronomy through the works of Nagasaki-based scholar Nishikawa Joken, who published popular books on the subject.[40] Yoshimune appointed Takebe as an advisor and joined the latter in making astronomical calculations. In 1720, perhaps urged on by his advisors, Yoshimune eased restrictions on the importation of Western books, allowing Japanese mathematicians access to astronomical literature written by Jesuits in China.[41] In 1726, Takebe learned that a new Chinese book had arrived in Japan, the *Encyclopedia of Astronomical Calculations* (*Lisuan quanshu*), written by a Chinese scholar of European sciences, Mei Wending, in 1723.[42] Following Takebe's suggestion, Yoshimune ordered another mathematician—Nakane Genkei—to study this treatise and render it readable for a Japanese audience by adding Japanese grammar marks (*kunten*) to the classical Chinese in which it was written.

These activities proved to be formative for Japanese astronomy. Yoshimune's lifting of the ban and Genkei's careful study of Mei Wending's

Encyclopedia of Astronomical Calculations led to increased demand for Chinese treatises on European astronomy. In the following decades numerous other books were imported from China, including the *Chongzhen Calendar* (*Chongzhen Lishu*),[43] compiled by Jesuit astronomers Giacomo Rho and Adam Schall von Bell; *Compendium of Observational Astronomy* (*Lixiang kaocheng*),[44] written by a student of Schall's; and the visually impressive *History of Instruments Used in the Observatory* (*Lingtai yixiang zhi*),[45] which described instruments found in the imperial observatory in Beijing that had been designed under the guidance of Jesuit astronomers. (see fig. 3.3).

Figure 3.3. An image of instruments in the Beijing Imperial observatory from the *History of Instruments Used in the Observatory* (*Lingtai yixiang zhi*). Waseda University Kotenseki Sogo Database of Chinese and Japanese Classics.
The image of the observation platform shows both novel Western instruments such as the quadrant, the sextant, and the celestial globe, but also the astronomical instruments that were already in wide use by Chinese astronomers, such as the armillary sphere, the variety of sundials, and the meridian device. Still the instruments depicted in *The History of Instruments Used in the Observatory* ignited the imaginations of Japanese astronomers.

The so-called Western methods of Japanese astronomers were therefore a hybrid, built through several layers of interpretation. Their interest in European trigonometry was inspired by the proto-trigonometry of the thirteenth-century Chinese *Granting the Seasons* treatise. The books they imported were already a hybrid, created in China by Jesuits and their Chinese students, in which European terms were replaced by classical Chinese astronomical terminology, and calculation methods interpreted through locally familiar practices and conceptual frameworks.[46] Moreover, they rendered Western practices into a textual format that made sense to Chinese and later Japanese readers. When these treatises arrived in Japan, they underwent yet another round of translation and interpretation. Even if written in familiar-looking characters, these texts still needed to be edited for grammatical reasons. Thus, by the time Chinese treatises detailing Western astronomical methods reached Japanese astronomers, they had already undergone a multilayered process of interpretation and integration into a familiar scholarly framework.

Yoshimune's Visions of Reform and Abe Yasukuni's Dawn-Observing Device

The importation of treatises on Western astronomy from China was supposed to facilitate a new calendrical reform that the eighth shogun, Yoshimune, had been planning since soon after taking office in 1716. The idea that there was a need for another such reform had been brewing since the turn of the century, promoted by parties with very different interests. On the one hand, there were the mathematicians, such as Takebe Katahiro and Nakane Genkei, who had studied the *Granting the Seasons* system and were dismayed by Shibukawa Shunkai's rounding up of observational data to fit his cosmological ideas of numerical harmony.[47] On the other hand, Yoshimune was interested in calendrical reform as a political tool. His appointment as shogun had been a matter of chance. He was no closer than a second cousin to the preceding shoguns and was nominated thanks to a series of unexpected deaths, including those of his two older brothers. He was thus seen as a relative outsider, despite being a member of the broader Tokugawa family. Upon assuming power, he envisioned a series of economic and institutional reforms, collectively known as the Kyōhō reforms,[48] that would position him as the Confucian ideal of a beneficent ruler, driven solely by his duty to improve the Japanese people's lives. Motivated by the idea that one of a ruler's responsibilities was to "grant the seasons" to the people, he saw complaints about discrepancies between the official calendar and observable phenomena as a stain on his rule.[49]

Yoshimune envisioned Western astronomical instruments as key to his planned reform. Before learning about Western astronomy he thought of reverting the calendar to the original *Granting the Seasons* system. He was advised against this by Takebe, who pointed out the advantages of Shunkai's correction of the system for the geographic location of Japan. Learning about Western astronomy from Nishikawa Joken, Yoshimune grew fond of European instruments, without necessarily understanding the mathematics associated with their use. Thus, for example, he ordered the installation of a large telescope at the central observatory, pointing it directly to the sun. The sun shone through the telescope on the surface below, which was painted with a grid. He then waited for the large mechanical tower clock to strike noon, and measured the position of the sunbeam on the grid.[50] Such observations probably produced little data of value, since for astronomical purposes it was important to know the exact timing of the local noon, and no mechanical clock at the time was reliable enough to tell that. Yet Yoshimune remained enthusiastic about foreign instruments that, for him, embodied the utility of Western sciences. Unlike mathematical calculations, the details of which only professional mathematicians of Takebe Katahiro and Nakane Genkei's rank could understand, instruments were material and clearly *did something*.

Yoshimune's preference of instruments over mathematics determined his choice among the candidates to reform the calendar. He preferred the popular writer Nishikawa Joken to his advisor, the mathematician Takebe Katahiro. Nevertheless, Joken's old age—he was sixty-eight when Yoshimune took office—and subsequent death in 1724 prevented him from leading the reform efforts. But his fame carried over to his son, Masayasu, who was instructed to continue his father's work and to design a new calendar according to the principles of Western astronomy.[51] "Principles of Western astronomy" had a very specific meaning for both Yoshimune and Masayasu; rather than using the calculation methods of Western astronomy, they focused on the most impressive and easy-to-grasp features they found in the newly imported books, namely the observational tools: sextants, telescopes, and, of course, the exemplars of cutting-edge machinery—mechanical clocks. The presence of the clock, for Yoshimune and Masayasu, had a value more symbolic than practical; the time that it measured was appreciated not for its end result but rather for the process of measuring itself—mechanical and fashionable.

Masayasu was assisted by Abe Yasukuni, a member of the Tsuchimikado family, who were traditionally responsible for the compilation of the yearly civil calendar.[52] The Tsuchimikados' role was not to come up with a better calendrical algorithm, nor to gather new data, but rather to use the existing algorithm to create a yearly schedule for the everyday use of the populace. Yasukuni, however, did not restrict himself to that traditional role. He extended his activities to astronomical observation and was in constant communication with Masayasu concerning the astronomical side of the calendar compilation process.

In making his celestial observations, Yasukuni also devised his own timekeeping devices. One of them was a *shaku* clock, similar to ones mentioned in the previous chapter. Named the "hundred-*koku* clock," or "seasons-granting pole,"[53] it referenced the *Granting the Seasons* system and Shibukawa Shunkai's adaptation of its instruments. Since Yasukuni's job was to translate astronomical time into civil time measured in variable hours, his clock had both the movable "civil" hours and the hundred equal astronomical *koku*.

But the device Yasukuni used the most was a particular type of sundial, described in his *New Book on Calendrical Methods* (*Rekihō Shinsho*).[54] This instrument seems to have been a hybrid of the "upward-facing instrument," described in the *Granting the Seasons* treatise, and the movable equatorial sundial designed by Shunkai. Honorifically called the Dawn-Observing Device (*Senshigi*), it was a tilted copper disc, with four rings inscribed on each side.[55] The first two rings on each side were used for the measurement of hours in equal units, while the remaining rings were made specifically to reflect variable hour distribution for different seasons. The upper side of the disc would be observed from the vernal equinox to the autumnal equinox, when the sun was to the south of the disc; the underside was used during the period when the sun shone from the north of the disc, from the autumnal to the vernal equinox.[56] By observing the elongated shadow of the pointing needle, Abe could know the time both in the variable units and in equal astronomical units during each of the twenty-four mini-seasons. This shadow-casting needle was carved in the middle to create a fissure, "thin like a line," which allowed only "droplets" of sunlight through. As a result, the shadow cast by the needle onto the ring had a thin, bright center, which, according to Yasukuni, provided an exact measurement of time, "without even a thin-hair-sized difference from the hours set by the Heavens"[57] (see fig. 3.4).

This design of this device reflected Yasukuni's use and conception of time measurement. His customary task as a Tsuchimikado had been to

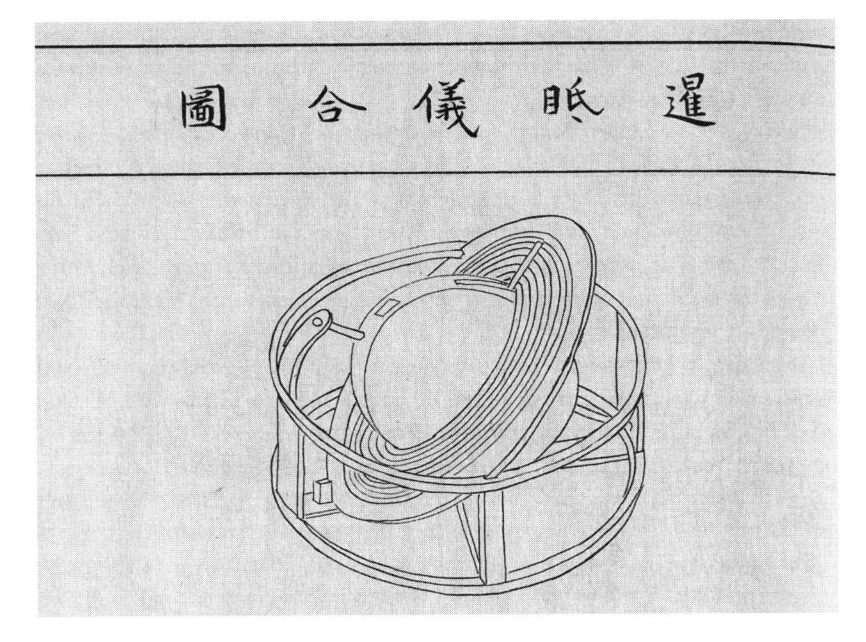

Figure 3.4. Abe Yasukuni's Dawn-Observing Device. National Diet Library Digital Collections. The concentric rings on this sundial correspond to the seasons. In order to know the time in variable hours, the observer would identify an appropriate ring and trace the movement of the shadow on that ring alone. Abe Yasukuni, *New Book on Calendrical Methods (Rekihō Shinsho)*.

translate astronomical algorithms into a calendrical schedule that could be understood and used by the broader audience. In this practice, he would take numerical values derived from astronomers' mathematical calculations, and round them up to match the gross values used in everyday life. His sundial mimicked the illustration of seasonal changes in the length of hours (see fig. 2.9) and provided time measurement in both sets of temporal units *simultaneously*, making his task of translation between the two systems virtually automatic.

Since Yasukuni was not an astronomer, his invention was actually not a measuring, but rather a notifying device, similar in function to the timepieces of everyday use. As its name suggests, the Dawn-Observing Device was supposed to indicate the timing of dawn—the beginning of the day. However, as noted earlier, since dawn happens gradually there was no observable phenomenon that determined that point. Rather it was determined mathematically, through astronomical calculations. Consequently, the Dawn-Observing Device could not be used to *determine* the time of the dawn through observation, but rather served to notify its user of an already predetermined point of time. Like Yoshimune and Masayasu, Abe was concerned not with acquiring new data, but with comparing the existing data to observable phenomena.

And since he was not concerned with acquiring new data, his requirement for time measurement precision was geared toward everyday, not astronomical use. Although he claimed that his device told time "without even a thin-hair-sized difference from the hours set by the Heavens," the smallest unit on the timepiece was one *koku*, or approximately fifteen minutes. As long as the timing fell within the range of this unit, Yasukuni considered it to be precise.

Yoshimune oversaw the enactment of many of his envisioned reforms, but the calendrical reform for which he had long hoped had still not taken place by the time of his abdication (1745) and death (1751). The reform was finally carried out in 1754, the fourth year of the Hōreki era, and it was universally considered a failure. Although during the first year of their use the new calculations seemed to fit the observable phenomena, by the second year the discrepancies were already larger than those of the previous calendars, and several years later the Hōreki calendarical algorithm was unofficially replaced by the so-called Shusei algorithm.[58]

The failure of the reform did not surprise astronomers and mathematicians. Even before its implementation, a lot of criticism had been directed toward both Masayasu and Yasukuni, who were deemed to be amateurs lacking in astronomical knowledge. As one critic put it, they "didn't study mathematics and could not make any calculations," and "even when [Masayasu] used astronomical devices, he didn't make sure to align them properly on the north-south line, or put them on a horizontal plain."[59] Their use of timepieces did not meet the standards of astronomers' time measurement, as they used "only one clock of a hundred *koku* alone, with no special person in charge of it," which "was sometimes ahead and sometimes behind."[60] The failure of their reform was thus seen as a direct result of their failure to treat time in the manner that astronomers did.

Arcs and Angles in the Kansei Reform

One person particularly critical of the Hōreki reform was Asada Gōryū, whose perceptions of time measurement were based on a different kind of astronomical practice.[61] In 1763 Gōryū predicted the coming of the solar eclipse that the Hōreki calendar missed completely, also correctly calculating its timing and magnitude (i.e., how much of the sun was covered and where). Gōryū was originally a private physician to the lord of the Kitsuki domain but in the course of his service he developed an interest in Western sciences—first anatomy and then astronomy.[62] Gōryū spent his spare time studying mathematical works by Takebe Katahiro and Nakane Genkei, and later the Jesuit treatises imported from China. Eventually, he decided that he wanted to pursue astronomy exclusively, and left Kitsuki.[63] By 1763, his astronomical expertise clearly surpassed that of the designers of the Hōreki reform.

Unlike Nishikawa Masayasu, Gōryū not only was swayed by the impressive instruments depicted in Chinese treatises but also learned a new way of thinking, pertinent to Western astronomy. In propagating Western science, Jesuits sought to convey Western concepts in language familiar to their Chinese audience. They thus worked with their students on writing their treatises in classical Chinese, utilizing the terminology of Chinese astronomy, and employing the format of Chinese astronomical treatises. Their attempts were so convincing that some Chinese observers declared that Western astronomy must have originated in China.[64] This familiar language, terminology, and format also played a crucial role in the acceptance of these treatises in Japan, allowing Japanese astronomers to focus on the content, and to make connections between novel Western practices and their own professional concerns. Thus, they enabled the gradual incorporation of Western modes of representation, calculation, observation, and even the very way of thinking about astronomical problems into Japanese astronomy.

Jesuit treatises did not actually teach the astronomy practiced in Europe at the time. Subjected to the orders of papacy, Jesuits were forbidden to teach the Copernican model of a heliocentric universe. Instead, they utilized Tycho Brahe's cosmological hybrid, in which the planets revolved around the sun, but the sun revolved around the earth. Seen as a gross misconception today, this model was actually fairly useful for calculating celestial events. More importantly, however, was the fact that cosmological questions were not the locus of the transformative influence these texts had on Japanese astronomy. Rather it was their mathematical methods, which necessitated a different kind of calculational practice and a different kind of time measurement.

The Jesuits' books taught how to precalculate celestial motion by using spherical trigonometry that relied on diagrams in which arcs represented measured time. Spherical geometry, and trigonometry in particular, required visualizing the problem by drawing such diagrams, in which angles were formed by connecting an observer's position on earth with two particular points on the path of a chosen celestial body. By measuring the time it took for the celestial body to get from one point to another, astronomers acquired necessary numerical data that could be later utilized in their calculations. With these kinds of calculations, rather than focusing on the timing of intervals between celestial events, as had been their common practice, Japanese astronomers shifted their focus to the motion of celestial bodies. By representing this motion in drawing, they began thinking of their numerical data—including time measurements—as most informative when embedded in a diagram. Since the paths of celestial motions they were measuring were represented as arcs in a diagram, the act of time measurement became associated with arcs and angles.

This new astronomical practice required a qualitatively different kind of time measurement. In the previous astronomical systems, including the majority of calculations in the *Granting the Seasons* system, time measurement produced data that was analyzed in order to detect patterns that occurred over extremely long periods of time. This data would then be divided by a number of years, months, or days in order to determine how patterns manifested in the short term. An example of such a calculation is the previously described Law of Vicissitudes, which allowed Shibukawa Shunkai to compute that each year was getting shorter by 0.000730485 of a day and round this value up to 1/10,000 of day, which was equivalent to one *byō* of time. For Shunkai, the division of a *koku* into tenths (*bu*) and thousandths (*byō*) was a purely mathematical construct derived from computation alone. There was no clock at the time capable of measuring such brief durations, but there was also no need to measure *bu* or *byō* since they were *derived* from computation. Jesuit astronomical treatises, on the other hand, extrapolated from data gathered through the observation of a preselected celestial body, the motion of which was defined as an arc and calculated using spherical trigonometry. When extrapolating from a single diagram, the precision of the calculation depended on a detailed measurement of the arc. Its length was determined by measuring the length of time it took for the selected body to travel from one point to another, which demanded the measurement of time in smaller and smaller units. Miniscule temporal units ceased to be mathematical constructs alone, and became a practical necessity.

In order to satisfy the need for minuscule temporal units, Asada Gōryū commissioned a customized *shaku* clock. He probably ordered a *shaku* clock because it was considered to be the latest timekeeping technology, and also because Abe Yasukuni had already used a *shaku* clock for his observations. But Gōryū's clock was different. When he finally received it, he described it as incomparable for astronomical observations, indispensable even when one was equipped with an armillary sphere. The clock was painted with black lacquer and the subdivisions were scratched on its surface using a thin compass needle. With its huge size, it featured an unusually large dial, making it physically possible to inscribe additional subdivisions and thus increasing its precision.[65] Yet within just a few years, Gōryū felt that the *shaku* clock no longer satisfied his need for precision.

The solution again came from something he read in a Jesuit text. At the very end of the fourth scroll of *History of Instruments Used in the Observatory*, which describes the instruments used in the imperial observatory in Beijing, there are several pages concerning a "suspended ball" that could be held and made to swing[66] (see fig. 3.5). The duration of the oscillations was constant and could be used to measure the duration of astronomical events. It is not clear when exactly Gōryū started using a pendulum, but in the inventory of the tools he used to observe the eclipse of 1789, there is a description of a "suspended ball" that makes 34,438 oscillations a day, or 95.6612 oscillations for every degree of the sun's apparent daily motion.[67]

The pendulum measured a different kind of time. First, it measured periods of time that were less than a third of a *byō*—a unit that theoretically represented one ten-thousandth of a day but so far could not be measured by any other existing timekeeper. Furthermore, this level of precision was different from that achieved with previous measurement techniques not only in degree but also in kind. Although it appears that both Shibukawa Shunkai and Asada Gōryū were measuring time, they were, in fact measuring different things. The units used by Shunkai measured enormous cycles of reoccurrences of celestial alignments, while the units used by Asada Gōryū measured the distances traveled by celestial bodies. The small temporal units in Shunkai's system were needed in order to find the moment when numerous temporal cycles converged. In such a system, a slight divergence in time measurement would become apparent only over a period of time. Such was the case with Shunkai's numerical abbreviations, which manifested in the calendar's divergence from observable phenomena only after several decades. And this was similarly the case with Abe Yasukuni's gross time measurements, which could still produce a calendar that worked for a year. Trigonometry, however, requires a different perception of measured time. Here time units mean distance, and with each diagram representing distance on an astronomical scale, even a few brief oscillations of the pendulum could translate into hundreds of miles. The difference in the kind of mathematical methods employed by astronomers also correlated with a difference in the intended use of measured temporal units, and a different image visually associated with the measurement of time.

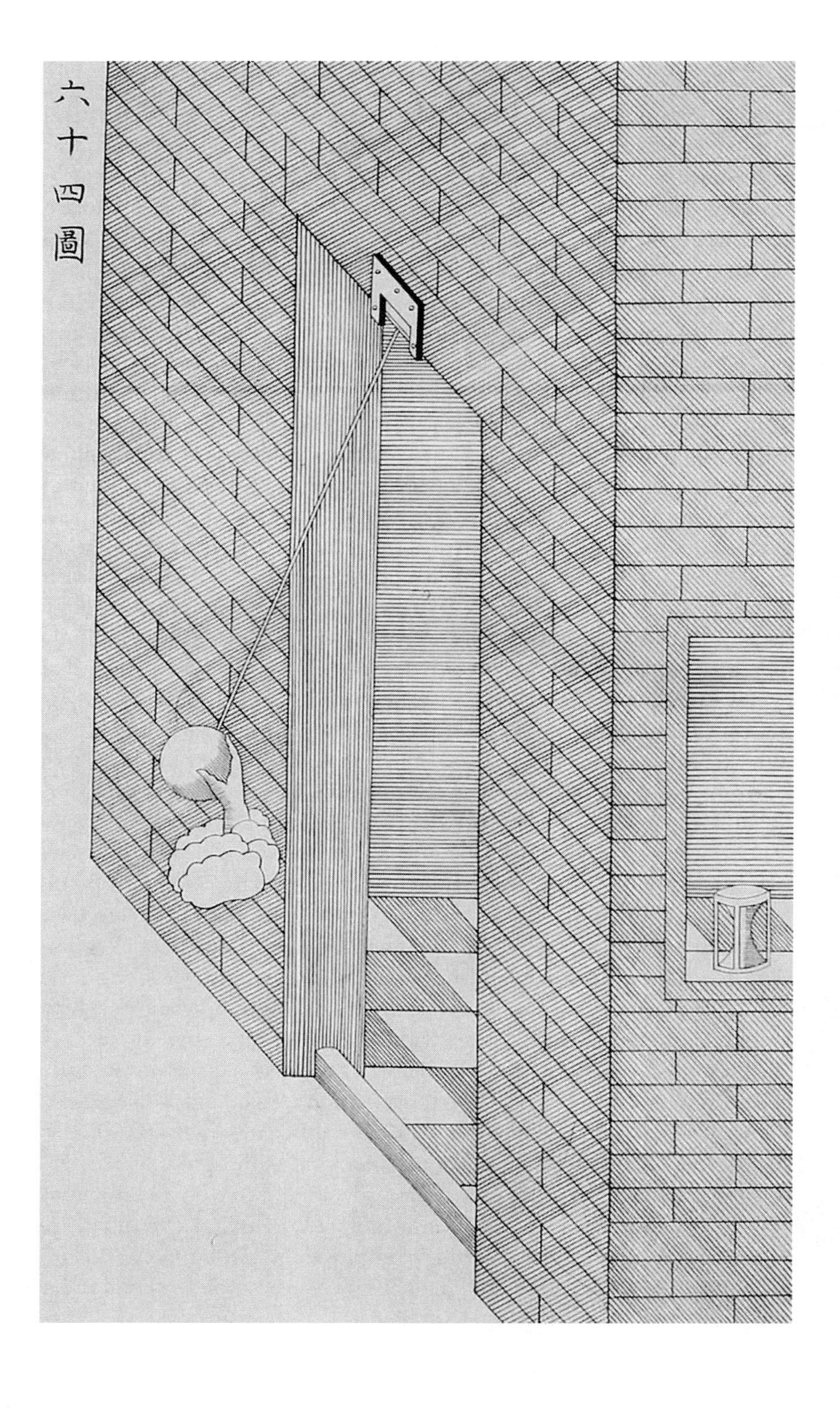

六十四圖

Minuscule units of time now were not only measurable, but also visible, defined by a little arc traversed by the pendulum.

This different perception of time measurement was incorporated in the next calendrical reform, in the Kansei period. The reform was initiated by Matsudaira Sadanobu, a powerful shogunal advisor, who de facto ruled the country around the end of the eighteenth century.[68] Matsudaira Sadanobu was a grandson of the previous great reformer, Yoshimune. Due to political rivalry he missed an opportunity to become a shogun himself, but through an adoption into the Matsudaira family he was able to gain a position as one of the "elders" at the age of twenty-eight.[69] Like his grandfather, he took rule of a country troubled by economic instability, droughts, and famines. And like his grandfather he envisioned a series of fiscal, institutional, and moral reforms collectively known as the Kansei reform.[70] An essential part of this project was a new reform of the calendar, and in 1795 Sadanobu summoned Asada Gōryū to design a new calendrical system. Under the pretense of old age and illness Gōryū politely declined, suggesting instead two of his favorite students—Hazama Shigetomi and Takahashi Yoshitoki.[71] Coming from strikingly different backgrounds, the two scholars were nevertheless close friends and complemented each other's abilities. Yoshitoki came from a poor samurai family, gaining an excellent education yet struggling to feed his family throughout his life; Shigetomi, on the other hand, was a son of a wealthy brewer, who had enough means to entrust the business to another party and dedicate himself to the study of astronomy and to his greatest hobby—the building of scientific instruments. Consequently Yoshitoki was always a man of theory and calculations, while Shigetomi was a master of measurement and observations. The two, assisted by a circle of astronomers, worked less than a year to devise the calendrical algorithm, which was implemented on the first day of the year Kansei 10 (1798).

As Takahashi Yoshitoki and Hazama Shigetomi were working on the new calendrical algorithm, they were introduced to the 1742 *Sequel to the Compendium of Observational Astronomy* (*Lixiang kaocheng houbian*).[72] Rather than relying on a Tychonian framework, *The Sequel* introduced Kepler's first

---

Figure 3.5. Illustration of a pendulum from *The History of Instruments Used in the Observatory* (*Lingtai yixiang zhi*). Waseda University Kotenseki Sogo Database of Chinese and Japanese Classics. The illustration depicts a simple pendulum attached to a sturdy horizontal surface and manually set in motion. The hourglass on the lower right side of the image suggests that the pendulum is a timekeeping device.

two laws—that the orbit of each planet around the sun is an ellipse, and that a line drawn between a planet and the sun sweeps equal areas during equal periods of time. According to the memoirs of Shigetomi's son, Asada Gōryū was so distressed after reading *The Sequel* that he wanted to burn all his own works and had to be physically restrained by his students to prevent him from doing so.[73] It was not the heliocentric cosmology that troubled Gōryū—Japanese astronomers saw the heliocentric model as a logical continuation of a Tychonian model and the shift did not require many calculational changes, as the relationship between the sun and the planets remained the same. The trouble was the elliptical orbit and the irregularity in apparent celestial motion, as it required reworking the whole calendrical algorithm. Even though the irregularity of celestial motion had been recorded as far back as the *Granting the Seasons* system, Kepler's laws put the uneven motion of the celestial bodies in the spotlight. Exploring the patterns of uneven motion provided an incentive to focus on segments of orbits, and measuring segments and arcs in units of time became essential for astronomical practices.

The new astronomical practices had a direct impact on the calendar. Unlike the former calendars that defined the times of dawn in terms of the number of *koku* units that had passed since midnight, the Kansei calendar defined the dawn in terms of a solar arc below the horizon. Specifically, the official moment of dawn—i.e., the beginning of the day—was defined as the instant when the sun was at 7 degrees (*do*), 21 minutes (*bu*), and 36 seconds (*byō*) below the horizon at the latitude of Kyoto[74] (see fig. 3.6). Since time was now seen as a reflection of celestial motion, the designers of the Kansei calendar strove to represent the fact that the motion of the earth around the sun—and hence the change of the seasons—happened at an uneven rate. Consequently, they determined that, contrary to the rule in previous calendrical systems, the length of hours should not change at the same rate throughout the year. In the Kansei calendar, the length of hours changed at a rate that reflected the earth's position in its elliptical orbit, and hence its speed, according to Kepler's second law. Additional innovation of this calendar derived from Asada Gōryū's version of the previously mentioned Law of Vicissitudes—a version that he had developed in his *Laws in Time* (*Jichūhō*).[75] According to Gōryū, and in contrast to the systems of both *Granting the Seasons* and Shibukawa Shunkai, the length of the year was not only diminishing but was actually based on a 25,400-year cycle.[76] Gōryū corrected the extremely large—by astronomical terms—values of yearly change that had been arrived at by Shunkai a century earlier, but in fact Gōryū's value, too, was extremely large and caused inconsistencies in

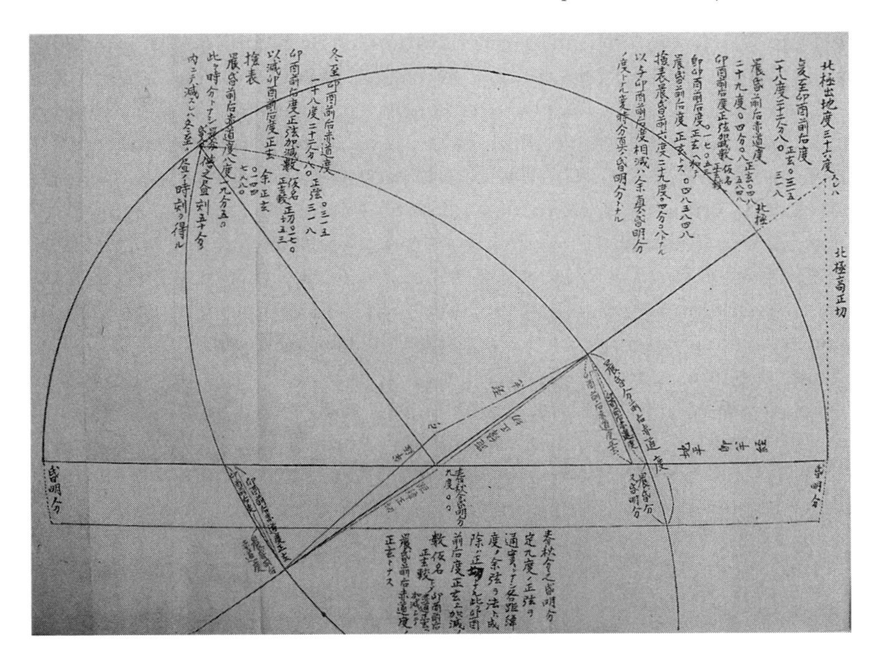

Figure 3.6. Inō Tadataka's *The Times of Days and Nights in Provinces and Districts* (*Kokugun chūya jikoku*). Japanese Academy of Science, Gakushiin Archive, Tokyo. This diagram was used for trigonometrical calculations of the moments that mark the beginning and ending of the day. The diagram reflects the novelty of the Kansei reform, which defined the beginning and the end of the day in terms of the sun's position below the horizon, measured in degrees, rather than in units of astronomical time, *koku*. The diagram provides both a visual and calculational rationale for treating time as an arc, rather than as a series of segments.

the new calendar. Thus, by the time the Kansei reform was completed it had become clear that the calendar would need to be revised yet again.

Pendulums and Celestial Motion

The new conceptualization of time measurement also resulted in the design of a new astronomical timepiece. Manual handling of the pendulum was tiresome, and could result in human-introduced errors. Recognizing this, Hazama Shigetomi began working on automating Gōryū's swinging pendulum.[77] According to his son's memoirs, Shigetomi "read in *The History of Instruments Used in the [Imperial] Observatory* explanations about the pendulum, grasped its principle, and after thinking it through, realized that there is regularity in the number of minute pendulum swings. By discovering the way to employ this natural movement by combining the hanging pendulum with mechanical gears, he created this device. He called it the Suspended Oscillating Disk Device [*Suiyōkyūgi*],[78] and with it he developed guidelines for a precise measurement of the time of celestial motions"[79] (see figs. 3.7 and 3.8). The device had separate dials for tens, hundreds, thousands, and tens of thousands of oscillations and counted roughly sixty thousand oscillations per day.[80] For the purposes of calculating segments of motion the particular number of oscillations did not really matter; as long as the oscillations were constant, they only had to be divided by 360 to reveal how many represented one degree of celestial motion. On the other hand, the function of "translating" the number of swings into temporal units meaningful to humans, which had been so central in the astronomy of Abe Yasukuni, was barely given any attention.

The time kept by the pendulum clock was quite different from any other temporal system in Japan before. Although the first device did have

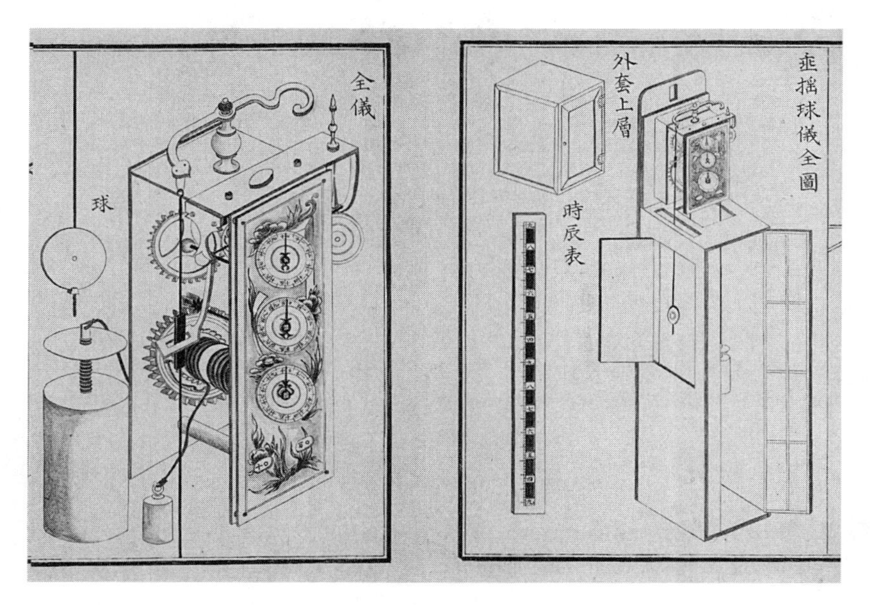

Figure 3.7. Suspended Oscillating Disk Device (*Suiyōkyūgi*) in *The Book of Kansei Calendar* (*Kansei Rekisho*),1859. National Diet Library Digital Collections. A detailed description of the device can be found in the book written by Takahashi Yoshitoki's youngest son, Shibukawa Kagesuke. In *The Book of Kansei Calendar*, Kagesuke detailed not only the practices and devices used to prepare the new calendrical algorithm, but also the ones that emerged later, as a result of new ways of conceptualizing time. While describing the pendulum clock, Kagesuke discussed its structure, its function, and its relative advantages.

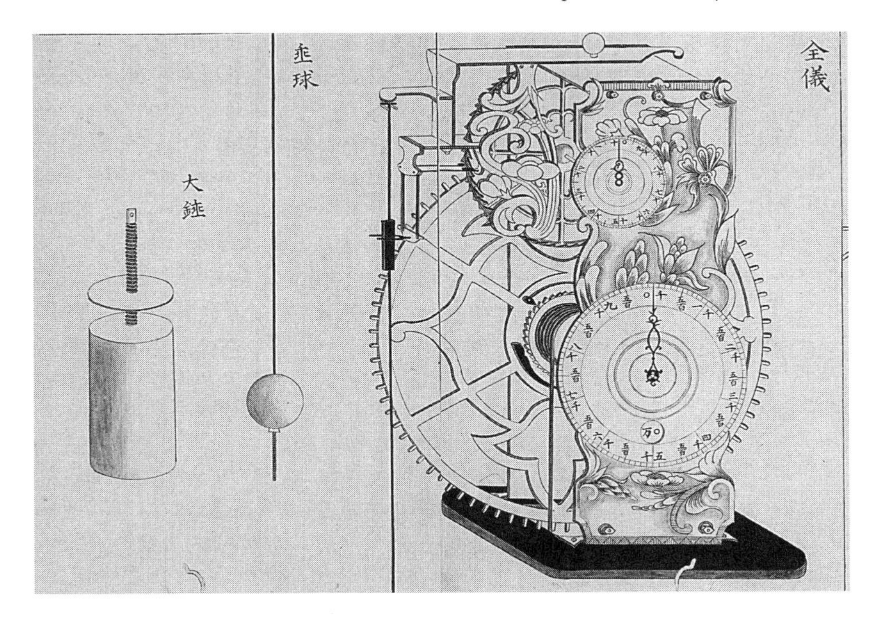

Figure 3.8. A later variant of a Suspended Oscillating Disk Device. *The Book of Kansei Calendar* (*Kansei Rekisho*). National Diet Library Digital Collections.

an attached *shaku* clock with hundred-*koku* astronomical units and everyday variable hours, it was hidden inside the box containing the pendulum, pointing to its lesser importance. Later designs of the pendulum clock got rid of the *shaku* clock altogether, which suggests that even its initial inclusion may have been done as a gesture to previous astronomical timepieces rather than out of practical consideration. The temporal units that did matter—the oscillations—did not fit any human standards. The pendulum clock did not measure portions of a day, because the new kind of calculations made by astronomers did not measure *events* that were defined in relation to specific dates. Rather, time as measured by the clock stood for the length of an arc created by a moving celestial body. For measuring the length of an arc, it did not matter what hour of the day it was, nor even what temporal unit was used for the measurement—what mattered was that the length of that unit was constant and minuscule.

Defined by the swing of the pendulum, the time of the Suspended Oscillating Disk Device was also associated with arcs. First, as we have already seen, there was the arc inscribed in a diagram. Since celestial motion represented by the arc was measured in time, by drawing the arc astronomers also drew a representation of time measurement. Second, there was an arc inscribed in the air by the oscillating pendulums. Each oscillation counted by the pendulum clock was visibly moving in an arclike motion, so that time itself manifested as a continuous series of arcs. Arcs were also inscribed by instruments used in conjunction with pendulum clocks, and even by the bodily movements of people who operated them. Time measurement with the pendulum clock required a collaborative effort, involving two additional instruments—the sextant (*shōgengi*) and the "meridian device" (*shigosengi*).[81] Thus, for example, an observer with a sextant would note an altitude of a star, shouting to the clock-keeper to start noting the time; a person operating the meridian device would notify the other two when the star crossed the celestial meridian, indicating the end of the measurement. Or the clock would be calibrated according to the local noon determined by the meridian device, which recorded the moment when the sun reached its zenith.[82] Since the pendulum clock was always used in conjunction with the meridian device and the sextant, the three were considered inseparable and often referred to in a single word, sextant-meridian-pendulum-device—*shōgenshigosensuiyōkyūgi*.[83] All three together were associated with arcs—the pendulum clock was following the arcs made by the swinging pendulum, the meridian device was marking the points on the arc inscribed by the moving celestial body, while the sextant was itself shaped like an arc and required the user to make an arclike motion while observing the moving celestial body with its telescope.

Reconstructing Western astronomical practices, Japanese astronomers also developed theoretical concerns reminiscent of their European colleagues'. Making calculations by drawing diagrams, Asada Gōryū began investigating the relationship between the physical structure of the universe and the timing of the movement of celestial bodies. Particularly, he was hoping to find a mathematical relationship between the positions of planets and their orbital periods. When, in the mid-1790s, Gōryū was introduced to the *Sequel to the Compendium of Observational Astronomy*, he learned about Kepler's first two laws, but *The Sequel* did not mention Kepler's third law, which provides a mathematical formula that describes the relationship between the periods of planets and their distance from the sun. Despite the fact that he did not learn about this law, Gōryū was preoccupied with exactly the same problem.[84]

Working with the pendulum, Gōryū's student Hazama Shigetomi also came to conclusions that resembled those of his European colleagues. While designing the new astronomical clock, Shigetomi began contemplating the meanings that the motion of the pendulum bore for astronomy.

In a small booklet titled *A Detailed Treatise on the Pendulum* (*Suikyūseigi*), he recorded his thoughts about the device. First, he described a particular experiment that he had performed in order to learn the attributes of the pendulum. He made five pendulums of different heights and let them swing freely, counting the number of swings. He then concluded that the motion of the pendulum was determined by the proportion between the length of its rod and the weight of its bob. Although this relationship is considered to be erroneous today (the weight of the bob does not affect the period), it nevertheless reveals Shigetomi's thought process and the role of his associations in the making of his theories. He claimed that the regularity he discovered in the motion of the pendulum had, in fact, a much broader and more substantial meaning, stating that the proportionality between the pendulum's dimensions and its period was in fact embedded in "mathematical regularity of natural principles that does not allow even the smallest deviation."[85]

With this statement, Shigetomi provided perhaps one of the earliest formulations in the Japanese language resembling the Western concept of a "law of nature." The word that he used—*shizen*—is usually considered to be a Meiji term, whereas in earlier, Edo-period thought, *shizen* simply meant "on itself." Shigetomi's use of this word, however—both in the context in which he employed it and the meaning he meant to convey—show that the conceptual transformation of this term began much earlier.[86] The meaning of the word *shizen* was transformed because European practices infused the term with new associations.

Relying on the notion that similar mathematical laws governed all the phenomena in the universe, Shigetomi claimed that the pendulum could be used to learn about planetary motion. "Consequently," he wrote, "one can use the pendulum to explore, and clearly demonstrate natural regularities."[87] Extrapolating from the principle he discovered, he stated that "celestial bodies have weight, and all the bodies that have weight follow the same principle as the one behind the [movement of the] pendulum."[88] Namely, "since the movement of the pendulum depends on the proportional relation of its weight, if we fathom the regularity of the pendulum, the regularity behind the natural movement of the heavens will instantaneously become clear, as if a lamp was lit in the midst of total darkness."[89] As such, he claimed that the motion of the pendulum could help clarify the relationship between the weights of the planets and their distance from the sun.

He suggested imagining the planets as a set of pendulums hanging from the center—the sun—in order to explain the differences in planetary motion. After describing how planets move around the sun like pendulums with rods of different lengths, he stated that "for this reason, the rotation of a planet close to the sun, such as Mercury, is quick, whereas the rotation of planets distant from the sun, like Saturn, is slow. The principle behind this proportion between the slow and quick [movements] and the distance of the planets from the sun is exactly the same principle that governs the [motion of the] pendulum"[90] (see fig. 3.9).

Shigetomi relied on the association between time and pendulum motion to perform a kind of thought experiment that would help him arrive at a theory of planetary motion. He found regularity in pendular motion and attempted to extrapolate from this regularity the positions and motions of the planets. His investigation of the pendulum was therefore not only an investigation of a mechanism, but also a means to the much deeper and broader understanding of *the* mechanism—the mechanism that governs the world—where things are "exactly in the place they are supposed to be."[91] His notion of time measurement, as associated with arcs, particularly of those of the pendulum, defined his investigation of the regularity behind celestial motion.

Shigetomi's musings about the pendulum bear remarkable similarities to several core assumptions of European thought—that there are mathematical regularities in nature, that similar regularities govern timepieces and planetary motions, and that we can learn about planetary motion by studying our timepieces. He arrived at these ideas by conducting pendulum experiments similar to those made by Western thinkers—Galileo, Huygens, and Newton among them. Yet it was not the pendulum itself that determined his conclusion. The pendulum only provided the material metaphor, while the interpretation of this metaphor was shaped by Shigetomi's calculational, observational, and measurement practices.[92] These practices were rooted in Western astronomy and formed core associations that channeled his interpretation of his experiments.

Shigetomi's conclusions were, of course, far from identical to those of his European colleagues. One striking difference was the absolute absence of any divine creator, which European thinkers deemed to be necessary to set the clockwork universe in motion. On the other hand, in explaining the distance of the planets from the sun, Shigetomi relied on the metaphysics of *yin* and *yang*. Such differences are not at all surprising. After all, Japanese astronomers were not *adopting* Western astronomy, nor did they conceive of themselves as *practicing* Western astronomy. Rather they developed their own astronomical system, which converged with several European astronomical principles due to the integration of Western astronomical methods into their practice.

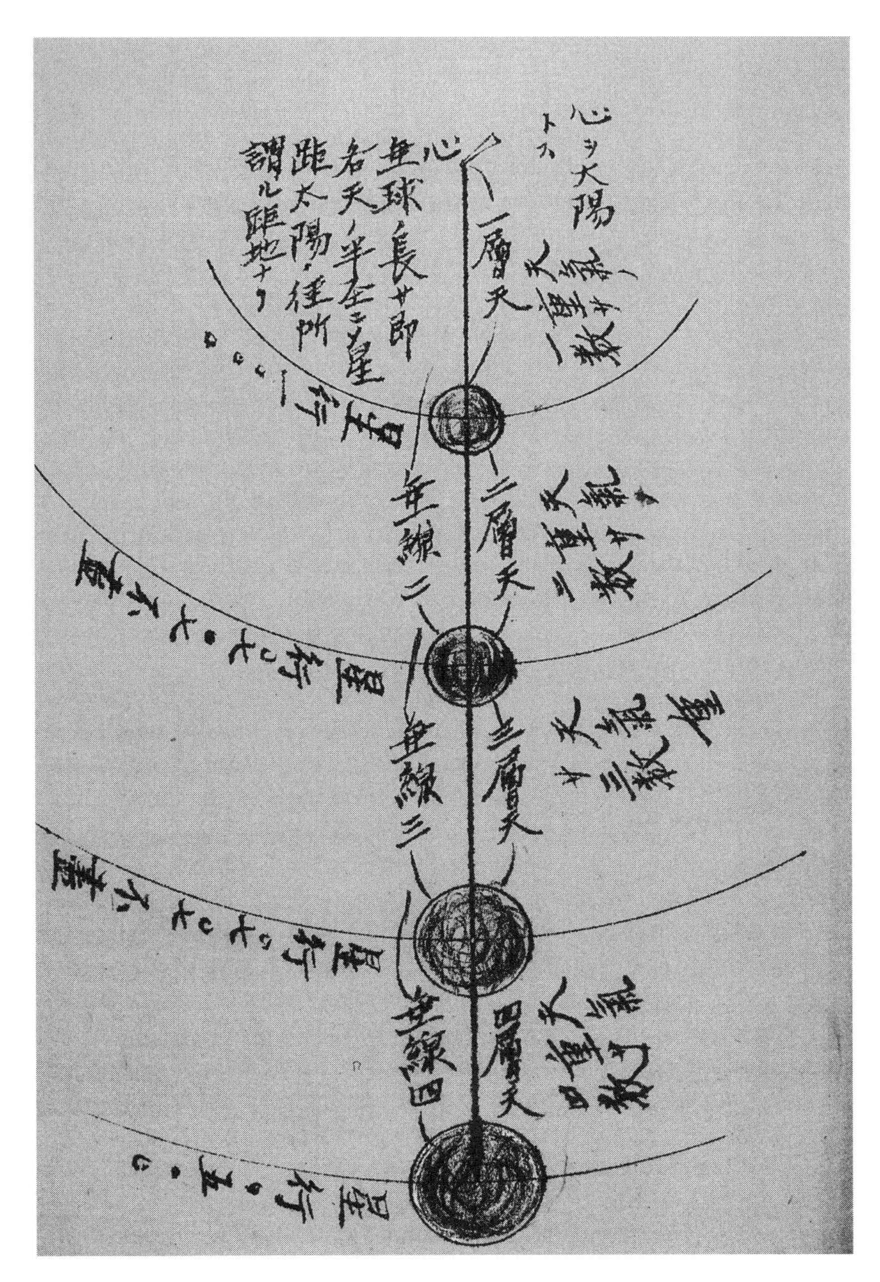

Figure 3.9. Pendulums as planets, *Detailed Treatise on the Pendulum* (*Suikyūseigi*), Japanese Academy of Science, Gakushiin, Tokyo. Illustration of one of Hazama Shigetomi's thought experiments, in which he imagined the planets to be pendulums hanging from the sun (labeled as "the heart" or "the center"). Shigetomi describes the motions of the planets on different "levels" in terms of the length of a pendulum rod and the period of its oscillation.

* * *

The very particular ways different astronomers used measured time shaped their associations with time measurement and thus their concept of time. Generally speaking, all astronomers observed and timed celestial phenomena. In principle, they all had the same goal—the study of the universe and the devising of mathematical algorithms that predict celestial motion. Yet the particular uses of measured time—what time stood for in mathematical calculations, what kind of temporal units were required, and how time was visualized—changed in the course of the seventeenth and eighteenth centuries, forming a new set of associations and transforming the overall conceptualization of time measurement. Time measured to represent universal harmony created different associations than time measured for trigonometric calculation. Time measurements written exclusively in tables bore different meanings than the ones represented as arcs in diagrams. What it meant to measure time changed with the first calendrical reform of the Tokugawa period, and continued changing until the Kansei reform at the end of the eighteenth century.

These meanings manifested in timepieces astronomers designed for themselves. The equatorial sundial of Shibukawa Shunkai reflected his conception of time as place-dependent. Nishikawa Masayasu's use of tower clocks reflected Tokugawa Yoshimune's enthusiasm about measurement instruments. Abe Yasukuni's sundial reflected his need to translate between astronomical units of *koku* and variable hours. And the pendulum clock reflected the spatial, spherical attitude toward astronomical calculations.

Western astronomy played a significant role in the transformation of the way time measurement was conceptualized. The growing interest in Western astronomy was not straightforward. The kind of Western astronomy seventeenth-century Japanese astronomers were exposed to was mostly cosmological, with few methodological benefits. And the only reason they wanted to learn more about Western mathematical methods was their similarity to those described in the thirteenth-century Chinese *Granting the Seasons* system. When they finally studied the methods of Western astronomy it was through Jesuit treatises written in classical Chinese. Yet those treatises transformed the calculational practices of Japanese astronomers, changing the ways they approached time measurements.

By learning about Western astronomical practices, Japanese astronomers transformed the world of their associations so that their conceptualization of time measurement began resembling that of European astronomers. No single person explicitly taught Japanese astronomers what their European colleagues thought about time measurement. Nor was this information contained explicitly in books. It was the immersion in the computational, visual, and material practices of European astronomy that provided Japanese astronomers with new keys for the interpretation of European technology. Just using Western astronomical instruments—as Nishikawa Masayasu did—could not transform one's whole perception of time. But by gradually integrating Western trigonometry methods into their practices, Japanese astronomers formed associations that allowed them to interpret European artifacts in a manner resembling the intent of the original makers. They learned the priorities of Western astronomy, they learned new methods with which to approach certain problems, and they learned how to work with artifacts central to European astronomy, which provided them with key material metaphors on which to base their speculations. By using similar methods and relying on similar associations to channel their interpretations, they increasingly arrived at conclusions similar to those of their European counterparts.

FOUR

Geodesy, Cartography, and Time Measurement

The seemingly esoteric practices of astronomers impelled a series of events that would eventually bring about a redefinition of the notion of geographical space. The wheels were set in motion by the calculation troubles of Takahashi Yoshitoki—one of the architects of the Kansei calendrical reform whom we met in the previous chapter. From the moment of its institution in 1795, Yoshitoki knew that the new calendrical system was far from perfect. His miscalculation of the time of an eclipse, which occurred almost seven minutes later than predicted, proved that his suspicions were well grounded.[1] And there were additional, less apparent, and less publicly embarrassing miscalculations involving the altitudes of culminating stars, which appeared to be slightly, yet constantly, off. He checked and rechecked his math but found no obvious computational failure, thus concluding that the problem must lie elsewhere. If the computation itself were correct, there could be two other potential sources of error: the translation of the raw numbers of celestial motion into actual places on earth, or the calendrical algorithm he was using to make predictions. Or both.

The former initially seemed the more probable source of error. There was persistent confusion concerning the actual length on the ground of one degree of the meridian. The various Jesuit astronomical treatises imported from China gave varying values, measured in traditional units of *ri*.[2] Some suggested one degree was 32 *ri*, others 30 or even 25. Even worse was the fact that it was not exactly clear how long one *ri* was. In Japan, one *ri* was equal to 12,960 *shaku*[3] but there were also different opinions about the length of one *shaku*, which could be, in modern-day terms, 29.6 cm, 30.258 cm, or 30.363 cm, and it was up to every artisan to determine his own standard for calibrating his tools.[4] But Yoshitoki also knew that the values of Chinese units of length were altogether different, and one Chinese *ri* (or *li*, in Chinese) was equal to only 1,500 *shaku* (or *chi*), also of undecided length.[5] Translating all of these uncertainties into modern terms, the length of one degree of the meridian could be as great as 126 kilometers (78.3 miles) or as small as 10 (6.2 miles).[6] Western books arriving in Dutch translation through Nagasaki only added to the confusion. Japanese interpreters tried to figure out the length of the *voet* (foot) in Amsterdam versus that of England, of France, or of "the lands of the Rhine."[7] Not to mention that the Dutch translations of French texts failed to mention the very recent adoption of a standard metric system in France, and mostly referred instead to the old French *toise* and *pied*.

The second possibility, that the calendrical algorithm itself was wrong, was hardly less likely, even if for the simple reason that astronomical algorithms are *always* wrong to some extent. With the tumultuous changes in Japanese calendrical theories and practices over the course of the eighteenth century, no astronomer could be naïve or optimistic enough to think that his model would be the one to capture all the parameters of celestial motion, and Yoshitoki in particular was keen to acknowledge his previous mistakes.

One reason for his doubts may have come from an outside source. Hungry for any information about Western astronomical practices, Yoshitoki was quite intrigued by the books that arrived at the VOC (Dutch East India Company) colony in Nagasaki. He could not read Dutch, and when he referred to Dutch-language books in his writings he often stated this or something similar: "I don't know Dutch and hence it is difficult [for me] to comprehend [the explanation]."[8] Nevertheless, he was proficient in another kind of language used in Dutch books to communicate ideas—the language of images, diagrams, and numerical tables. Educated to a large extent by Chinese-language Jesuit treatises, Yoshitoki developed a habit of not only reading geometrical diagrams and numerical tables, but also thinking through drawing and analyzing information by organizing it in a table. Consequently—as he himself stated—when looking at Western books he was reading images and attempting to derive rules based on his understanding of them.[9] At some point, he encountered an entry titled "The World Globe" in Egbert Buys's 1769 *Woordenboek*, to which he was probably referred by a scholar of so-called Dutch studies. The encounter with that particular entry was perhaps largely fortuitous, as the Dutch for "world" is "Aard" and thus conveniently located at the very beginning of the first volume of Buys's huge, multivolume encyclopedic dictionary.[10] As usual, Yoshitoki couldn't understand the details of the nine pages of text, but there was a table that caught his eye.[11]

Even without a proper understanding of Dutch, Yoshitoki noticed that the table held potentially grave implications for his calculations. He could recognize the words *Myl* (mile) and *Graad* (degree), and he could certainly read Roman numerals and also likely recognized the ° symbol. Thus, even without comprehending the text, he could see that different degrees corresponded to different numerical values—that the length along the surface of the earth of one degree of meridian varied according to its location relatively to the equator and the poles (see fig. 4.1). If the Dutch book was right, the earth was not a perfect sphere, meaning all previous calculations that had been based on the assumption that the globe was uniformly curved were incorrect. As he explained it several years later, previous calculations assumed that if one described observation angles in a diagram, and then drew a line perpendicular to the observation point, this line would point to the center of the earth. However, if the earth was not a perfect sphere then this line would go elsewhere. Consequently, astronomical calculations based on the assumption that the earth is a perfect sphere would not be borne out by actual measurement.

The Dutch table confused Yoshitoki. Besides the fact that he didn't know how to translate Dutch miles into Japanese units, he was also reluctant to accept apparently bizarre Western values for the lengths of the meridian at different degrees of latitude. Buys's table stated that the length of one meridian degree at the equator was 60 miles, which then drastically dropped by half a mile at 10°. Afterward, however, the tendency reversed

AARD. 35

Breedte.	0.	10°.	20°.	30°.	40°.	50°.
Myl in een Graad.	60.	59,5.	59,57.	59,67.	59,8.	59,93.
Breedte.	60°.	70°.	80°.	90°.		
Myl in een Graad.	60,6.	60,16.	60,235.	60,26.		

Figure 4.1. The entry for "Earth" ("Aard") in Buys Egbert's *Nieuw en volkomen Woordenboek van Konsten en Weetenschappen* (*A New and Complete Dictionary of Terms of Arts and Sciences*), Vol. 1, 1769. "Breedte." is short for "Breedtegraad"—latitude. "Myl in een Graad" means "miles in one degree." Even with basic knowledge of the Dutch language Yoshitoki could read this table and decipher its meaning. According to the table, the length of one meridian degree at 0° latitude (the equator) is 60 miles. This value then drops to 59.5 miles at 10° latitude but grows until it again reaches 60 miles at about 57° latitude. When translated into an actual shape, these values result in an oblate ellipsoid with a bulge along its long axis.

and one meridian degree measured longer and longer as the latitudes climbed toward the poles. If the table were correct, the earth would have a bulge around the equator.[12] For Yoshitoki, this didn't make any sense. Several years later, he would say simply that the table had a mistake in it, but when he initially encountered it he was just perplexed.[13] He could not take any foreign sources for granted, and had to find the way to make his own tables and determine the actual length of one meridian degree at the latitude of Japan.

Measuring One Degree of the Meridian

A solution for the problem of the length of the meridian degree was offered by one of Takahashi Yoshitoki's students—Inō Tadataka.[14] Inō was an unusual student. Born into a high-class yet very poor family, he later married a widowed daughter of a wealthy brewer and was given the task of improving the finances of the brewing business. He was quite successful, gradually gaining not only wealth but also a high-ranking position in local administration. He traveled extensively on business between his native village of Sawara and Edo (Tokyo). Interested in the terrain he saw on his travels, he decided to survey it using techniques he learned from the available surveying guides.[15] His life changed in 1795 when, at the age of fifty, he decided to leave his business and the management of the village to his eldest son and travel to Edo to study astronomy under Yoshitoki, twenty years his junior. Owing to his wealth and age, exceptional for a student, Inō received more attention than usual and proceeded quickly through the regular curriculum. He began with Shibukawa Shunkai's *Jōkyō Calendar*, the thirteenth-century Chinese *Granting the Seasons* calendarical system, and the mathematical works of Takebe Katahiro, which Yoshitoki considered the basis of any astronomical course of study, and only then moved on to Jesuit astronomical treatises imported from China.

Inō's surveying aspirations provided Yoshitoki with an exceptional opportunity to resolve the problem of the length of one meridian degree through a measurement of geographic distance alongside one degree of latitude. At first, Inō measured both the geographic distance and the latitudinal difference between his native Sawara and the Astronomical Bureau in Edo, but the distance was short and Yoshitoki feared that any minuscule inaccuracies in its measurements would amount to a considerable degree of error in any calculation based on them.[16] In 1800, Inō made a longer surveying journey, taking a path leading north from Edo. Upon his return, he reported that the length of one meridian degree north of Edo was 27 *ri*.

Yoshitoki was fairly satisfied with this value, as it matched both his own estimations and the values in his trusted Chinese treatises. This could well have been the end of the story. It was not.

Although Yoshtoki and Inō both had good reasons to be content with Ino's results, each nevertheless felt that the survey had not fulfilled its purpose. Inō did not have the chance to fully employ his recently acquired knowledge of astronomical observation techniques and equipment, while Yoshitoki felt he had missed the opportunity to obtain additional data from observations taken in locations far from the Astronomical Bureau in Edo. Thus, in spite of their general satisfaction with the value of one meridian degree Inō had obtained by surveying, the two started making plans for another expedition to the north.

The second expedition differed in its scope and routine.[17] In his earlier survey, Inō had employed techniques he had learned from the existing literature on surveying—he used a compass to determine directions and measured distances by using ropes and chains, or a perambulator (when conditions permitted), or by pacing with an odometer. In 1801, on the other hand, he planned to make astronomical observations alongside his surveying activities. He thus equipped himself with the latest astronomical instruments—a variety of theodolites, sextants, and most importantly, a pendulum clock.[18]

Inō's new instruments were specially ordered from the well-known Edo instrument maker Ōno Yasaburō.[19] Ōno was one of two famous clockmakers who worked with astronomers to design new instruments for astronomical calculation (the other, Toda Tōsaburo,[20] worked with astronomers based in the Kyoto/Osaka area). Ōno's role in the mission cannot be underestimated. Since all the instruments were calibrated to match his specific measuring tools, for the sake of consistency Inō also decided to adopt Ōno's value for the length of one *shaku*. The value, which still appears in modern dictionary definitions of this ancient unit, was set at 30.34 cm. As it was standardized by Inō's measurements, it came to be known as the *Inō-shaku*.[21]

The sheer number of heavy instruments Inō intended to carry as well as the complexity of his planned observations (some of which required several participants) turned the intended project into a large and expensive expedition, which required official sponsorship. Yoshitoki and other influential acquaintances of Inō petitioned the government, but officials hesitated, doubtful about the practical value of measuring the length of one meridian degree. To make the mission more appealing to them, Inō promised to extend his survey as far as the northern island of Ezo (modern-day Hokkaidō), which was still considered "barbarian" territory. Ezo was rich in natural resources and hence of particular interest to the central, shogunal government. Nevertheless, this cunning suggestion backfired when the government officials demanded that the expedition take a boat to the northern end of the main island, which would drastically reduce the cost but also undermine the entire observational project. Calculations of cost and benefit were the government's primary concern, and the astronomically verified value for the length of a meridian degree did not seem, at that point, to be worth the investment. Seeing that the mission was on the verge of being cancelled, Inō, who had already subsidized his studies from his own pocket and spent a considerable amount of money on customized measurement instruments, offered to fund most of the expedition, even though he had to cut the number of people who accompanied him. The government accepted the offer, providing him an official stamp for the mission and a modest stipend. Given Inō's own financial resources, the stipend was not the most crucial factor to support the project.[22] More significant was the government seal that granted the expedition party priority on the roads and the promise of cooperation from local inns.[23] Accompanied by his surveying assistants, Inō finally set out on his trip in May 1801. Taking measurements along the way, the expedition headed north through the main island of Honshu, crossed the sea to reach the northern island of Ezo, and traversed its southern coast before returning to Edo.

To Inō's great disappointment, Yoshitoki refused to accept the new value he brought back with him—28.2 *ri* (110.85 km).[24] In a letter to his friend and fellow astronomer Hazama Shigetomi, Yoshitoki expressed doubts about Inō's astronomical competence, arguing that carrying out observations and making calculations was a very difficult task "even for a superior surveyor" like Inō. Examining all the values known to him, Yoshitoki conjectured that the length of one meridian degree (from Edo to Utsunomiya) could not possibly be larger than 27.5 *ri*.[25] Inō was bereft. In his letter to Yoshitoki's son and student Takahashi Kageyasu he lamented: "I spent three years chasing the value of one meridian degree, determining it to be 28.2 *ri* but [Yoshitoki[26]] doesn't believe me, because it does not match the numbers known so far. . . . Even though I prepared all the calculations and observations for him to see, he still doesn't believe me! He wouldn't comfort my old heart. Why does [he] doubt me so much? . . . If he doesn't believe in me to such a degree, maybe I should just stop the surveying of the distant places from now on!"[27]

Things changed, however, in the beginning of 1803, when Yoshitoki received a copy of a Dutch translation of Lalande's *Astronomie.* As previously described, his Dutch was very limited and he relied instead on his understanding of the graphs and tables in the book. Through his interpretation of nonverbal textual elements, he was able to form an understanding of geodesy—the measurement of the earth—on the basis of which he developed a new algorithm for calculating the length of one meridian degree for the latitude of Japan.[28] His new value stood at exactly 28.27326 *ri*, extremely close to the one measured by the triumphant Inō.[29]

Although the original goal was achieved, Inō continued his surveying, which gradually developed into a grandiose project to map the whole of Japan. Yoshitoki did not get the chance to enjoy the newly found confidence in the value of one meridian degree and the propagation of Lalande's astronomical theories in Japan. In early 1804, he fell ill and passed away at the age of forty-one. Inō's expeditions, on the other hand, received more and more government attention and funding. Officials were still unable to grasp the importance of his astronomical findings, which had benefited only a handful of professional astronomers. However, they could not fail to notice potential uses for his highly detailed land surveys, especially of the previously undersurveyed island of Ezo, over which the Japanese government was trying to tighten its control.[30] During the following years, Inō and his team set out on more and more expeditions, surveying coastlines, highways, and mountains. With each mission, the government bestowed on him greater privileges and bigger budgets. Before his death in 1818, Inō's group completed another ten major expeditions and numerous local surveys, crisscrossing all four main islands of Japan.

Land Surveying Using Time Measurement

It should be noted that the changes in the government's attitude were not motivated by the specific results produced by Inō's expeditions, as he rarely brought back more than raw data, in which officials had little interest. The promise of future benefits lay not in the expeditions' results, but rather in their methods of astronomical surveying, which were substantially different than anything previously undertaken in Japan (see fig. 4.2).

Much of what we know of the expeditions' methods comes from the extensive surveying diary Inō kept, the surviving parts of which amount to several modern-day volumes.[31] Specific surveying techniques are described in his numerous other writings, as well as in short treatises that his students wrote about his methods.[32] Looking at the described techniques, we

Figure 4.2. *Land Surveying* by Katsushika Hokusai (1848).
The Sumida Hokusai Museum, Tokyo.
The popular image of land surveying was shaped by what was mostly visible from outside—the laying of ropes, pacing with odometers and perambulators, and measurements with theodolites. The invisible tasks of surveyors that involved the actual compilation of data remained obscure until it was projected on a map.

can see that in terms of land surveying per se, there was nothing in what Inō did that had not been used in Japan before: the exploration of the terrain was mostly achieved using the traverse and the triangulation methods of land surveying, while the actual measurements of length were accomplished with a perambulator (when conditions permitted), by pacing with an odometer, or by simply laying ropes and chains.

The innovation of Inō's expeditions lay in his reliance on astronomical practices, and particularly astronomical timekeeping. His acquisition of a pendulum clock—of the same design as the Suspended Oscillating Disk Device invented by Hazama Shigetomi and used by Takahashi Yoshitoki—transformed his geographical pursuits.[33] In his *Addendum to the Map Concerning the Number of Leagues from the Eastern Capital*, Inō explicitly stated that what made his 1801 expedition different from the previous ones were his nocturnal observations, made with the help of a sextant, a meridian device, and the pendulum clock.[34] At the time, Takahashi Yoshitoki praised the pendulum clock as the best possible device for making observations, indispensable for such practices as determining the longitude.[35] It is thus not surprising that he instructed Inō to carry such a clock across difficult terrain despite its weight and size, and record the times measured during observations[36] (see color plate 6).

Astronomical time measurement with the pendulum clock required consistency, which was based on the clock's calibration according to a local noon, at the moment when the sun crossed the meridian at the given location. Setting a pendulum clock according to the local noon thus became an essential part of a surveyor's routine. An expedition team surveyed by day and made celestial observations by night, setting up observation stations in the inns where they stayed (see fig. 4.3). In the course of his eighteen-year surveying career, Inō recorded data from more than twelve hundred observation stations. In twenty of these he made measurements over the course of several days to correct observational errors. Given the fact that Inō omitted from his records those stations that did not produce sufficient data due to inadequate meteorological conditions, we can assume that there were actually many additional stations at which he determined the local noon and set his pendulum clock in order to time celestial phenomena.

In this regard, the surveyors' routes were marked with observation points, which were defined in terms of local time. The practice of astronomical time measurement, which became the major component of Inō's

Figure 4.3. *Middle-of-Night Observation* (*Yonaka sokuryō no zu*) in *Illustrations of Urashima Surveying* (*Urashima sokuryō no zu*). National Diet Library, Tokyo. The illustration depicts nightly observations during Inō's surveying of the Kure area in 1806. The large surveying party, with their peculiar practices and astronomical equipment, made a big impression on witnesses. A local artist was moved to create a whole scroll of paintings depicting various stages of Inō's surveying process. The artist depicted only those instruments that were placed outside, to allow direct observation, and thus omitted the pendulum clock, which was probably kept indoors nearby. According to Nakamura Tsukō, the person depicted using the transit instrument (*shi-go sen gi*) is probably Inō himself.

expeditions, and which distinguished him from previous surveyors, suggested that all the different locations on their routes had a certain relationship to one another—a relationship that could be defined in terms of the pendulum time. Various locations were no longer dispersed destination points in the surveyed area, but rather a chain, an interconnected whole.

This conceptual linking by mode of time measurement was particularly apparent in the case of longitude observations. The idea of calculating longitudinal values using pendulum clock measurements had emerged almost accidentally. In 1798 Hazama Shigetomi happened to be back in Osaka when a meteor was seen in the skies of Japan. Needless to say, both he and Takahashi Yoshitoki observed the phenomenon, making simultaneous measurements in Osaka and in Edo. The experience of simultaneous measurements inspired a sudden realization about which Shigetomi wrote to Yoshitoki: "Asada Ryūtatsu[37] mentioned that by making a simultaneous observations of a meteor in different places it might be possible to calculate the distance between them."[38] Yoshitoki replied that this was indeed so, and that not only meteors, but any kind of eclipse would produce a calculation of distance when simultaneously measured with the pendulum. A month later, he wrote to Shigetomi again, providing him with calculations of the distances between Edo, Osaka, and Kyoto that he had arrived at through the comparison of observation times of the meteor.[39]

Yoshitoki consequently instructed Inō to make observations during his surveys for the purpose of determining the longitude—albeit with a slight change. Although the observation of an eclipse provided the optimal comparison, eclipses, even lunar, were quite rare and frequently disrupted by bad weather.[40] Therefore, Inō mostly observed the occultation of the satellites of Jupiter—a method that Yoshitoki had probably encountered in one of the Chinese-language Jesuit treatises.[41] Sometimes Inō compared his measurements to the tables prepared beforehand by Yoshitoki, and sometimes observations were made simultaneously at wherever Inō's party happened to be located and at the observatory in Edo, and compared later.

Through this comparison, clocks created a conceptual link, a link of time, between different observation points—between an unknown locality and the central observatory in Edo. At that point, there was still no map that reflected surveyors' data. The relationship between the central observatory and the various locales was established on the basis of raw numbers of pendulum oscillations. In many cases longitude could be determined only after an a posteriori comparison with observational data from Edo. Consequently, there was no map carried from the unknown observation locations to the center, Edo. Yet even before the surveyors' measurements were written down on a map, even before the archipelago came to be visually represented as a geographic unity positioned on a grid of longitude and latitude lines, in the surveyors' minds the various regions they traversed came to be conceptually unified as one continuous space linked together by the oscillations of pendulums.[42]

The importance of astronomical time measurement to the conceptualization of the surveyed lands can be clearly seen in the exploration of the north. Although Inō did cross over to the northern island of Ezo in his first expedition, he surveyed only its southeastern coastline. Moreover, to cross the strait to Ezo, Inō's party had to leave behind its heavy equipment—the largest theodolite and the pendulum clock. This segment of the surveying expedition was thus the only one that was not accompanied by time measurement, which had a serious impact on the accuracy of its geographical measurements. Although his maps of the rest of the Japanese archipelago are extremely close in accuracy to modern-day ones, Inō's map of Ezo appears to be at an angle compared to more recent maps of Hokkaidō.

Inō was not alone in his mapping aspirations. In the course of one of his expeditions to the north he met an amateur surveyor who asked to be taught some surveying techniques. The amateur was Mamiya Rinzō, who, after spending some time with Inō and his team, set out on his own, further exploration of Ezo.[43] He sailed north, far beyond where Inō had gone, finally establishing that the land known to the Japanese as Karafuto was the same as what Russians called Sakhalin, and that the body of water to its north was not a large river but rather a strait, meaning that Karafuto was not a peninsula but an island.[44]

Unlike Inō's expedition, however, Mamiya Rinzō's survey was not astronomical. From the diary of the Russian captain Vasiliy Golovnin, we learn that Mamiya's knowledge of astronomical measurements was no more than rudimentary.[45] According to Golovnin, Mamiya had astronomical tables but only knew how to determine latitude by the less accurate method of measuring the sun's altitude. Regarding the question of longitude, Mamiya demanded that Golovnin teach him the method of lunar distances.[46] But such a calculation demanded observational practices that were simply impossible for a lone traveler making his way on a light boat. He definitely could not conduct observations similar to those of Inō's party. The sheer weight and size of a pendulum clock alone made it hard to travel with, it was impossible to operate on a boat, and it was useless without a sextant and a meridian device, which had to be operated simultaneously—thus requiring the participation of multiple observers.

Consequently, the lands Mamiya surveyed remained in the category of the distant and alien north, albeit ones that held fewer secrets than they had previously. Mamiya did not determine longitudes on his own, and in search of latitudes extrapolated from the values of meridian degrees as determined by Inō. Estimating latitude by this method was far from satisfactory, although it did achieve the goal of linking surveying materials to the data acquired by Inō through astronomical observations. By integrating Ezo and Sakhalin into the coordinates established by Inō, Mamiya created a superficial impression of the continuity of the map.

Cartography as the Inscription of Time

It took years before raw numbers were made into maps. Although Inō was gradually amassing more and more data on the various regions of Japan, the maps produced in the process were fragmented and inconsistent. Takahashi Yoshitoki's eldest son, Takahashi Kageyasu, who continued his father's astronomical investigations after the former's death in 1804 and later succeeded him in the Astronomical Bureau, had published maps based on Inō's measurements in 1809.[47] But they were on a very small scale and could not possibly represent the richness and the accuracy of the measurements made by Inō. When the enterprise of projecting all the data acquired through terrestrial observations and celestial measurements on a proper scale finally began in 1818, Inō Tadataka was already seventy-three years of age, and the project was far from complete when he passed away the following March. Having been involved in this mapping project from his childhood, and more or less serving as Inō's right-hand man since Takahashi Yoshitoki's death, Takahashi Kageyasu became the chief architect of the map of Japan.

When the project was completed in 1821, it yielded maps that revolutionized the perception of Japan's space. The final product consisted of three sets of maps: a large map consisting of 214 sheets of regional maps, a middle-sized map of eight sheets, and a small map of the whole of Japan in three sheets. The large map, covering the area from the southern island of Tanegashima to the northern tip of Ezo, was on the scale of 1:36,000, the middle-sized one was on the scale of 1:216,000, and the smallest was 1:432,000. It was not only the scale of these elaborated maps that proved to be innovative. They presented a whole new genre (see color plate 7).

Previous maps had been made for specific purposes, for specific goals to be accomplished in the field—planning of infrastructure projects, calculating taxation, delineating boundaries between domains or private properties, and providing orientation for tourists and pilgrims. Previous maps' structures, modes of representation, and foci all differed, being adapted to their particular purpose. Thus, maps intended for infrastructural purposes focused on the dimensions and variations of features such as river basins; maps made for taxation purposes delineated borders and the size of crop fields; maps for tourists and pilgrims pointed out major landmarks and described routes from the perspective of a traveler. Depending on the purpose of the map, actual relative dimensions were often included only for the focal points, while irrelevant details were frequently omitted.

Inō's maps were different. An individual map's reference points were not significant landmarks on the earth, but rather celestial phenomena observable from all the different points on the map. The format of the map was not determined by its purpose, but by the mode of its production—astronomical measurements. All the observational data was mapped on latitudes and longitudes. Unlike previous Japanese maps, this grid was not borrowed from maps made by Europeans, but determined on the ground by Japanese surveyors. The astronomically determined grid was integrated into material characteristics of the land, and manifested in the map.

More importantly, the central meridian was set to run through Kyoto—the traditional cultural capital of Japan (see fig. 4.4). This choice was not obvious, since it was Edo, not Kyoto, that hosted the central observatory.[48] Thus, the data that Inō collected was compared to the data produced at the longitude and latitude of Edo, not Kyoto. But Kyoto was considered to be the cultural heart of Japan and official calendars distributed to the whole of the country were based on its local time.

This culturally driven choice produced an important political effect. The process of comparison between data gathered locally and at the center of calculation is said to be empowering for the latter.[49] However, in this case, the data gathered from many different locales empowered a location that was culturally rather than observationally significant. By orienting the map around the central meridian of Kyoto, all the distant regions of the Japanese archipelago became conceptually tied to the country's cultural center. These distant locales came to be defined relative to Kyoto's longitude, and the time measurement that was essential to the determination of longitude created one inseparable continuity—Greater Japan.[50]

The new map provided a novel conceptual tool for political validation. During the Tokugawa period, Japan was divided into a large number of semi-autonomous domains whose governance was delicately balanced between the central and the provincial governments.[51] Consequently, previous maps, even when they attempted to show all the territories, emphasized the borders separating domains. The new maps, however, not only

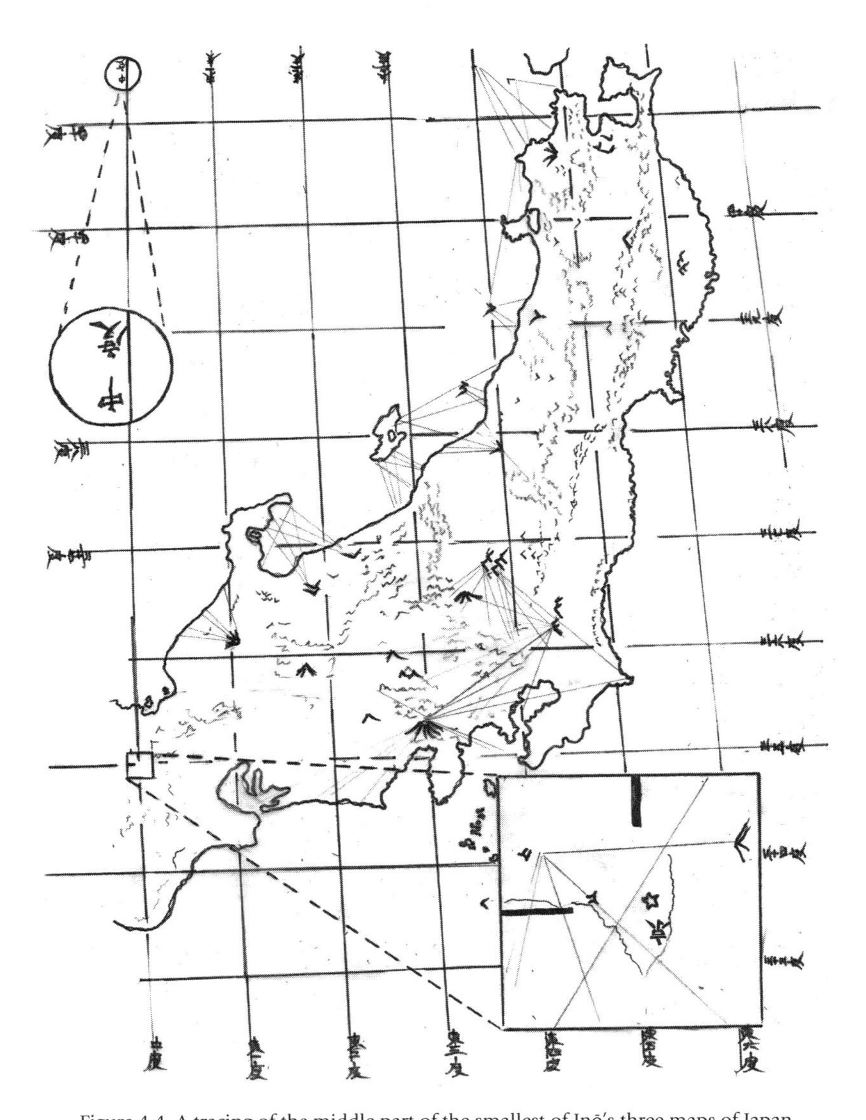

Figure 4.4. A tracing of the middle part of the smallest of Inō's three maps of Japan (refer to color plate 7 for original). *Comprehensive map of the coasts and the lands of the Greater Japan. Small Map. Eastern Part of Honshū* (*Dai Nihon enkai yochi zenzu. Shōzu. Honshū tōbu*). Original size 245 × 164 cm. Tokyo Metropolitan Library. The map is projected onto a grid of longitude and latitude lines that Inō determined using celestial observation. The longitude lines are labeled at the top of the map, with the prime meridian labeled "central degree"(中度 in the callout). To the right of the prime meridian the longitudes are labeled 1East, 2East, 3East, etc. The latitude lines are labeled along the map's side. The first from the bottom is labeled 33, the second is 34, the third is 35, etc. The intersection of the prime meridian and the 35th parallel (in the square callout) is located in the heart of Kyoto, indicated by a little star and by the character 京 (written upside down).

provided an overview of distant regions, but also represented a whole, a unified entity that was defined not only by political contracts between various domains, but also by observed celestial phenomena. Now the unified political entity derived legitimacy not only from mere human agreement, but also from the physical structure of the universe itself.

The pendulum clock, which produced measurements for the more accurate surveying of the land, transformed the very way space was perceived and viewed. The maps of 1821 were a consequence (and, at least in part, an unintended result) of Inō's initial expedition, which hadn't necessarily been motivated by nationalist aspirations. Yet the incorporation of timekeeping practices into surveying served as a platform for the development of the idea of geographic and national unity.

The Siebold Affair

For several individuals the new political perception of unified Japanese space bore grave consequences. On the night of September 17, 1828, a powerful typhoon struck the southern city of Nagasaki, wrecking the residence of Philipp Franz von Siebold. Siebold was a young German physician at the VOC colony on the artificial island of Dejima. Siebold, a naturalist, had sought employment with the VOC in order to explore the peculiar flora and fauna of the Japanese archipelago. The typhoon badly damaged the *Cornelius Houtman*, the ship that was about to take him back to Europe. The ship lost its anchor and was wrecked on the nearby coast, prompting a rescue operation by the local Japanese government. Taking this as an opportunity to inspect the Dutch cargo, officials found numerous contraband goods. Among Siebold's possessions they found something they deemed to belong to the realm of national defense rather than natural history—copies of Inō Tadataka's detailed maps of Japan. The maps were confiscated and Siebold was immediately imprisoned and later banished from Japan. A quick investigation revealed that the map was given to Siebold by none other than Takahashi Kageyasu, the main figure behind the publication of the maps based on Inō's measurements.

Kageyasu was not a traitor, but rather an inspired thinker starved for information. In the early 1800s, he had become fascinated with the world outside of Japan. He was inspired by the expeditions that had extended all the way to the northern frontier of Ezo and beyond, and by world maps, such as John Blaeu's *Atlas*, that arrived in Japan through Nagasaki. In the years of the so-called isolation policy, Japanese scholars were, of course, prevented from traveling outside the country to pursue their subjects of inquiry. Yet this practical obstacle did not prevent them from exploring the world in their imagination, on the basis of second-hand reports and their newly acquired knowledge of geography. Northern territories—especially the Russian side of the continent—were particularly intriguing. The knowledge that Mamiya Rinzō had obtained in Sakhalin augmented reports by Japanese castaways who had spent time in Russia, as well as those of prominent Russians such as Laxman, Rezanov, and later Golovnin.[52]

Attempting to create a coherent picture out of these fragmented bits of information, Takahashi Kageyasu had become involved in numerous mapping projects. He explored the northern territories, contributed to the reproduction of an annotated map of the "Russian New Capital of St. Petersburg,"[53] and worked on several detailed world maps[54] (see fig. 5.2 in the next chapter). These various pursuits enabled him to fit together pieces of the geographic puzzle.[55]

Given these interests, it is not surprising that Kageyasu sought out foreigners, especially those who possessed at least some degree of academic knowledge. He was especially thrilled to meet young Philipp Franz von Siebold, who held extensive audiences with local scholars on his annual trips to the capital and maintained correspondences with many of them. The two men shared a desire for knowledge: for Siebold, about Japanese flora and fauna, as well as the country's costumes and history; for Kageyasu, about astronomy first and foremost, but also about anything related to the world outside of Japan. Though it is impossible to know the precise content of their communications, Siebold and Kageyasu exchanged books, specimens, and other objects of interest.

The new interest in geography was by no meams limited to a small group of scholars. The government, which had been initially skeptical of Inō's ambitious surveying plans, gradually came to see the mapping mission as a matter of territorial importance. It jealously held on to the fruits of surveying projects, restricting the use of Inō's maps to the shogunal castle alone, which only reinforced the new conceptualization of Japan as an inherently unified political entity. To protect its interests the government employed a network of spies. Soon after Mamiya Rinzō's solo expedition to the far north, rumors started spreading that he was motivated not only by the desire to explore distant lands but—in particular concerning his dealings with Inō's group—also by his supposed employment as a secret government agent. In spite of the rumors, Mamiya maintained many connections in the scholarly world, which allowed him to learn about Kageyasu's communication with Siebold.

Mamiya promptly reported this communication to the government,[56] leading to heightened official interest in the nature of Kageyasu's activities and relationships with foreigners. Nevertheless, government agents had no legal authority to search Siebold's residence in Dejima and had no proof of any wrongdoing. The typhoon and consequent rescue operation provided a perfect excuse to inspect all the items onboard and thus recover the materials Kageyasu had handed to Siebold.

By this point, Inō's maps were no longer just the end product of a project that had started as an astronomical surveying mission; as representations of a unified Greater Japan, they were closely guarded by the central government. No matter what motivations Kageyasu may have had, his actions were deemed treasonous. The fact that he held a senior rank in the Astronomical Bureau did not help; he was thrown into prison and brutally interrogated. Prison records state that he died shortly after "from a sudden and grave disease."[57]

Kageyasu's death did not end the ordeal. Since his trial had not yet ended, the authorities decided to preserve his corpse so that they could officially deliver the verdict to his dead body. Due to his relatively high social and professional status it was suggested that his body be preserved in sugar, but this plan was subsequently abandoned in favor of a better preserving substance—salt. Surviving records detail the gruesome procedure of emptying the corpse of entrails, stuffing it with salt, and storing it in a huge salt-filled barrel with an opening at the bottom to allow the escape of decomposition liquids. The procedure was evidently no less shocking then than it is for a modern reader, as shaken witnesses described how by the time the verdict was supposed to delivered, the lifeless shell that used to be Kageyasu looked like "a dried fish."[58]

According to the verdict, the gravity of Kageyasu's offense was directly related to the newly acquired meaning of the maps he had been responsible for bringing to publication. He gave away "maps that showed the surveying of Japan and Ezo," specifically "the surveying of the various regions from Kyūshū, to Ogura, and Shimonoseki." These maps not only "indicated all the place names in abbreviation" and "offered insights into all the [local] novelties" but also "specified the two degrees [of longitude and latitude]." In the eye of the government, Kageyasu was not merely sharing data, he was betraying "the whole of Japan."[59]

Such offence bore severe consequences. Kageyasu's already dead body was sentenced to death. His family and students were banished to remote islands, at least temporarily. Even beyond his immediate circle, all scholars working on European sciences came under suspicion and had to restrict their activities in order to avoid his fate. Siebold was humiliatingly banished from Japan, leaving behind a wife and a daughter whom he would be able to see again only some forty years later.

Yet the government did not succeed in restricting access to the information contained in Inō's maps. When he departed Japan, Siebold took with him many notes, specimens, students' reports, and copies of maps he obtained from Kageyasu, which he had managed to hide. After his return to Europe, he published a series of bestsellers—the *Fauna Japonica*, the *Flora Japonica*, and his magnum opus, the *Nippon*, which incorporated his personal diary into a general description of the Japanese people as well as their customs and history. In a pictorial addendum to the book, Siebold included the maps of Japan, which he attributed to Takahashi Kageyasu.[60]

The publication of the maps was an important event in European cartography. The exact shape of the Japanese coastline was not yet known in Europe, and the harsh northern areas of Ezo and Sakhalin had been a mystery. Neither the shape nor the position of Ezo had been settled, and it was not even finally determined whether it was an island or if it was connected to Sakhalin. Moreover, it hadn't even been known for sure whether Sakhalin itself was an island or just a long land bridge that started just across the strait from the main island of Japan and stretched all the way to eastern Russian.[61] Several Europeans had previously tried to determine the shape and position of Sakhalin—the French Jean-François de La Pérouse in 1787, the English William Robert Broughton in 1796–1797. In 1804 the Russian Adam Krusenstern surveyed the Western coast of Japan and named the waters that he sailed the "Japan Sea." His compatriot Vasiliy Golovnin had been on a surveying mission in 1811 when he was captured by Japanese authorities. These attempts, however, had all been made from ships at a distance from the shore, thus rendering their measurements far from accurate. Moreover, each one of these surveyors had made only partial measurements, which contributed only a small piece of the puzzle. The map given to Siebold by Takahashi Kageyasu, on the other hand, provided a unified view of a continuous landform combined with a detailed outline of the coastline and precise measurements of longitudes and latitudes. Krusenstern (who had discounted Japanese scientific knowledge in 1804)[62] came to praise the "progress Japanese sciences have made" in the precision of their coastal surveys and longitude and latitude measurements.[63]

Siebold, however, did not publish Kageyasu's maps unaltered. Rather, he placed coordinates acquired in the European surveys on top of the Japanese measurements and mentioned place names bestowed by the foreigners alongside the Japanese and the local Ainu ones. Siebold pointed out to his readers that Japan had its own system of coordinates, centered on

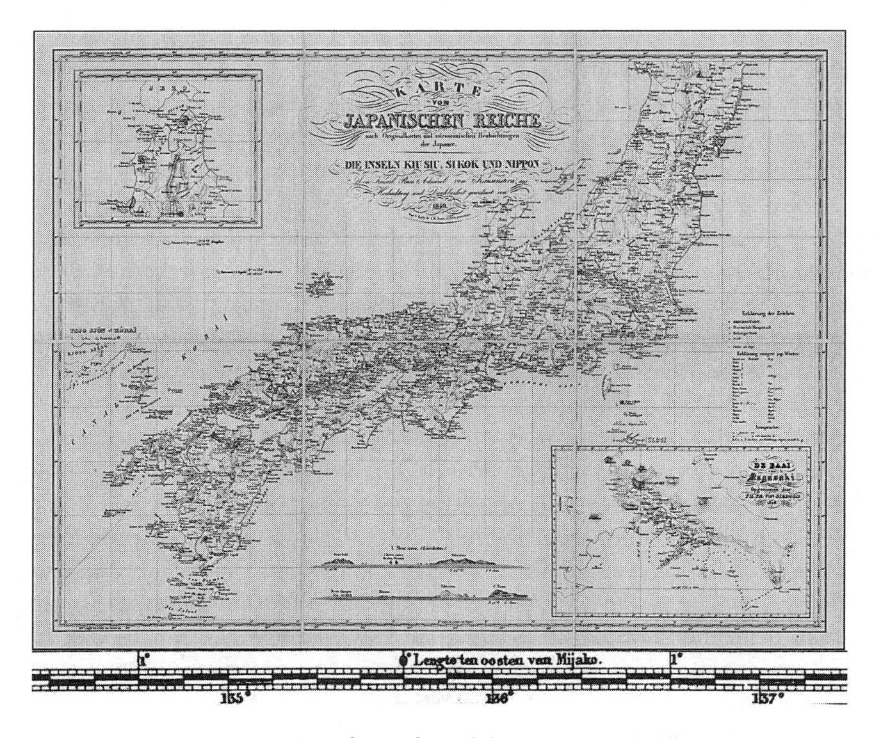

Figure 4.5. Map of Japan from Philip Franz von Siebold's *Nippon, Archiv zur Beschreibung von Japan.* The inscription reads "Map of Japanese States, from the original map and astronomical observations of the Japanese. The islands of Kiu Siu [Kyūshū], Sikok [Shikoku] and Nippon [Siebold probably confuses 'Nippon' for the name of the main island, Honshū]. Dedicated to the Imperial Russian Admiral von Krusenstern, with esteem and gratitude." The key at the bottom gives the longitude according to two scales. One is the "Greenwich-oriented Longitude" while the other is the "Mijako-oriented Longitude." "Mijako," or *Miyako* in the modern spelling, is an alternative name for Kyoto that means "the capital." On Siebold's map, Kyoto lies exactly in the intersection of the 35th parallel and the (Miyako-oriented) prime meridian, or 135°40′ from Greenwich.

Kyoto. But from the European perspective, Kyoto now was far away from the center. If, previously, Inō's map represented a temporal whole anchored to the culturally significant local time of Japan's historical capital, Siebold's new map positioned Japan as a distant periphery in the global flow of time[64] (see fig. 4.5).

* * *

When Takahashi Yoshitoki started having doubts about the length of one meridian degree, nobody saw it as a step toward a shift in the conceptualization of national space. His doubts were inspired by discrepancies between the predicted and observed times of celestial events, and his solution was based on a more nuanced and localized measurement of time. But the astronomical surveying project born of these doubts instigated new motivations, set new goals, and broadened the range of material associations with timekeeping to include surveying.[65]

The pendulum clock provided a standardized basis for comparison in the form of the period of pendular oscillations—a comparison that linked the simultaneous practices of various observers. Thus, even before the adoption of mean time and the Western temporal system, and even before astronomers themselves started using the Western chronometers associated with mean time, the pendulum clock came to be interpreted by Japanese astronomers as a device that provided material means to understand astronomical observations and geographical space in terms of time.

Astronomical time measurement thus had a unifying effect. It created conceptual links between different locations—links between observation stations along surveyors' routes, as well as links between remote places and the central observatory in Edo. It tied the earth with the heavens, distant places with Jupiter's satellites. It created the notion that the land was unified and defined by celestial motion.

Anchoring time measurement in the local time of Kyoto came to play an important role in the discourse of national unity. Far-flung places, including territories that could hardly be defined as "Japanese" at the time, were now all tied to the central meridian, and thus to the cultural and temporal heart of Japan in Kyoto. The maps that Takahashi Kageyasu based on Inō Tadataka's surveying expeditions were not only detailed charts of the Japanese archipelago, they were a material manifestation of a nation.

The events leading to the so-called Siebold affair thus represent a process in which a seemingly isolated practice was borrowed and transplanted from one area to another, leading to broader conceptual and practical changes. With the time-measurement practices Japanese geographers borrowed from astronomy came particular ideas about time and timekeeping. Therefore, the very data they acquired embedded astronomers' assumptions about time. This process was not intentional and the associations were not explicitly stated, yet they eventually had a profound influence not only on the active participants in the process of astronomical time measurement but also on the consumers of the end product, the map.

FIVE

Navigation and Global Time

Astronomical Data and the Problem of Temporal Units

Japanese astronomers initially turned to Western temporal units as part of a search for better data. They did not think that there was anything inherently better about Western timekeeping; nor were they motivated by social changes or broader power struggles. Rather, they started using Western units because vast amounts of astronomical data appeared in tables in European books imported directly from Nagasaki, which had not yet been translated into either Japanese or Chinese.

The need for European data originally derived from a specific problem that faced Takahashi Yoshitoki at the close of the eighteenth century. Even after the Kansei calendrical reform was implemented in 1796, Yoshitoki and his peers had continued to defend their teacher Asada Gōryū's Law of Vicissitudes, according to which the length of the solar year, which had previously been considered a constant, was determined to be gradually changing.[1] Defending Gōryū's theory was especially hard for Yoshitoki, since the algorithm that incorporated the projected rate of change produced a number of incorrect predictions.[2] But his belief in Gōryū's basic reasoning was firm, and he was determined to use empirical observations to find the true rate of this change. In order to do so, Yoshitoki required a much larger volume of observational data than was available to him in Japanese or Jesuit Chinese writings. This search for data led him to books brought by the Hollanders.

The fact that the data in Dutch books was recorded in Western temporal units did not pose a problem. Japanese astronomers were not at all troubled by the notion of alternative ways of measuring time. After all, they were already used to juggling several different systems of temporal units, including the Western units they found in Jesuit writings. For them, Western temporal units were just one of many possible mathematical expressions of time. The celebrated invention and use of a pendulum clock that did not operate according to any defined temporal structure exemplifies this attitude—for astronomers, the importance of temporal units lay not in their specific lengths but rather in the fact that they were constant and miniscule.

However, in order to make exact distinctions between units of different systems, astronomers needed to resort to novel forms of linguistic juggling. In earlier translations, Western units—minutes, seconds, and degrees—were glossed with existing Japanese terms—分(pronounced *bu* in the Tokugawa period), 秒 (*byō*), 度 (*do*).[3] By the end of the eighteenth century, however, astronomers had learned that although the two sets of terms occupied similar semantic fields, their precise meanings diverged. Take for example degrees. The numerical value of one degree derives from the notion that a circle represents an annual cycle of the sun—that is, a year. European tradition relied on an ancient abbreviation of the number of days in a year to 360. In Chinese astronomy, however, a similar tradition defined the number of degrees in a circle as 365, while another version insisted even more precisely on 365.25. Using the character 度 therefore required a specification of what system was this value referring too. Similar problems existed when talking about Western temporal units. In the astronomical system defined by Shibukawa Shunkai (based on the *Season Granting* system), a *bu* was one-tenth of a *koku* (which was one-hundredth of a day). One *byō* was one tenth of a *bu*. When Western "minutes" and "seconds" were referred to as *bu* and *byō*, they denoted different lengths of time altogether. Therefore, in letters written during the late eighteenth and early nineteenth centuries, astronomers constantly added notes indicating whether they were referring to a system that divided the day into 10,000 units (subdivisions of astronomical *koku*) or 1440 units (the number of Western minutes in 24 hours), or whether they were simply referring to a pure count of pendulum oscillations.[4]

The confusion between units prompted a broader search for standards, but it was not a foregone conclusion that such standards would be composed of Western units. Rather, this was part of a broader search for standards visible across early nineteenth-century Japanese society. This was the moment when Inō Tadataka was trying to determine, once and for all, the length of one *shaku* that could serve as a basic unit of measurement in his surveying. It was also the moment when the interpreter Baba Sajurō was trying to make sense of inconsistent information about Western units of measurement.[5] Prompted by nativist sentiments, Hazama Shigetomi too conducted a historical study of Japanese measurements, which has unfortunately been lost to history. While these attempts all sought to precisely define measures and units, they did not immediately indicate a preference for one kind of unit over another.

Yet, as Japanese astronomers began engaging with navigation, they came to see Western temporal units as qualitatively different from and preferable to the Japanese ones. Tokugawa astronomers' engagement with open-sea navigation may appear surprising, since up until the very last years of the Tokugawa period Japanese citizens were not allowed to sail abroad, while those who left were told to never return. Yet astronomers still discussed navigation on a mathematical and theoretical level and a favorable view of Western timekeeping emerged when they *imagined* what it would be like to navigate on the open sea. This view materialized not out of practical concerns but out of the series of new connections these scholars made between numbers, images, timekeeping devices, and navigational aspirations.[6] The fact that these practices took place in people's minds (and on the pages of the documents they left us), does not make them less significant. The hypothetical had a real impact on actual life.

Honda Toshiaki and the Beginning of Hypothetical Navigation

The first scholar to discuss open-sea navigation at length was Honda Toshiaki.[7] A political and economic theorist, Toshiaki sought a remedy for the ailing Japanese economy of his time. Writing in 1790, Toshiaki reflected on the string of disasters that had characterized the Tenmei period (1781–1789) of the previous decade. At the beginning of the Tenmei period, crops had failed after a string of cold winters. By 1783, with the grain reserves almost depleted, an additional disaster struck. After three months of preliminary eruptions, Mt. Asama exploded in streams of lava, pyroclastic flows, and clouds of ash.[8] Long after the initial disaster, the ash continued to suffocate the fertile Shinano plain and block sunlight to most of central Japan. The next few years produced few crops and resulted in one of the most devastating famines in Japanese history. In Toshiaki's view, however, the volcanic eruption was only an indirect cause: the real fault lay in the disastrous economic policies of the previous government.

In search of a good economic model, Toshiaki turned to Europe, which he imagined as a kind of utopia of creativity and efficiency.[9] He did not, of course, have any firsthand knowledge of life in Europe, but gathered information from stories told by the Dutch, as well as sporadic reports of Japanese castaways. Thus for example, he learned a great deal about Russia from Daikokuya Kōdayū, a sailor who spent years in the court of Catherine the Great. Kōdayū was brought back to Japan in 1792 by Adam Laxman, who falsely hoped that this gracious act would help to build trade between Russia and Japan. The notes and accompanying images from Kōdayū's subsequent interrogation inspired a fascination with Russia—and with Europe in general—after they were made public in 1794.[10] But the sporadic pieces of knowledge that Toshiaki gathered resulted in a rather naïve, if not plainly mistaken, vision of Europe.[11]

A consideration of what was actually happening in Europe during the 1790s makes his *Tales of the West* a particularly ironic read.[12] In Toshiaki's Europe there were thousands of ships filled with goods cramming the ports, universities that promoted only the talented, government institutions dedicated to justice, and officials attuned to the needs of the people. When Europeans got sick, Toshiaki claimed, they went to hospitals sponsored by the government where they were treated by the best possible doctors, and received a small stipend upon their recovery, to keep them afloat until they found a new job. In his description, all European countries exhibited nothing but excellence, efficiency, and foresight.[13] Their economies thrived because the Western mode of governance was based on the "three most important areas of study": astronomy, geography, and navigation.[14]

What made Western economies successful in Toshiaki's eyes was their governments' ability to counteract unexpected disasters by conducting international trade. Consequently, as early as the end of the eighteenth century, he advocated that Japan should lift its ban on international travel and begin to engage in commerce.[15] He believed that Japan was destined to conduct international trade because of its geographic location.[16] Although he was never directly exposed to European philosophy, Toshiaki made claims similar to those of European Enlightenment thinkers such as Montesquieu, insisting on a correlation between a country's climate and its style of government.[17] Toshiaki claimed that in order to learn the political potential of each country, it was essential to determine its exact location on the grid of latitude and longitude and thereby deduce its character-determining climate.[18] Thus, a given geographic location was not just a random place occupied by this or that country. Rather, it had an essential quality—its position on the earth relative to the stars shaped the character and the culture of the people who lived there. His comparison of Japan to another island nation, Great Britain (itself indicative of how little he knew about the actual climate there), claimed that since the two were geographically similar, Japan should emulate Britain's political and economic imperialism, occupy the northern island of Ezo, and later conquer Kamchatka.

Repeatedly stressing the importance of "astronomy, geography, and navigation" in his writings, Toshiaki decided to learn astronomy from Takahashi Yoshitoki. In addition to his official astronomy students, Yoshitoki also communicated with a broad array of public intellectuals, government officials, and wealthy amateurs interested in astronomical and geographic practices. By the turn of the century, the rise of utilitarian philosophy (*jitsugaku*) had led numerous high-ranking officials to take up observations and measurement as a form of leisure. Many of them wrote to Yoshitoki to ask his advice about astronomical practices or the proper operation of the expensive instruments they had acquired.[19] Honda Toshiaki, however, was more than just a lay technology enthusiast. He had a specific goal in mind—to learn how to sail abroad.

At the time, Yoshitoki was studying data from the British Nautical Almanac of 1795. Nautical almanacs were hefty books that provided precalculated data concerning hourly changes in the positions of celestial bodies throughout the year, allowing navigators to determine their position at sea. They were a relatively recent development in Western navigational practice, having been published only since 1776. Although Yoshitoki's original interest was the astronomical data contained within the 1795 almanac, the fact that it was a navigational tool inspired him to think more broadly about the mathematical methods entailed in navigation.

In 1803 Yoshitoki summarized his thoughts on mathematical practices in navigation in *Thoughts about Seafaring Routes* (*Kaichu funamichi kō*). The original work was lost in the fire that destroyed the Asakusa observatory in 1810 but was rewritten by Yoshitoki's youngest son, who claimed to have reconstructed the text from memory.[20] The surviving work is a short treatise on spherical trigonometry, analyzing the calculation of the shortest routes on the surface of the globe. By rooting his mathematical calculations in shipping routes, Yoshitoki contributed to the increasingly widespread imagination of navigation practiced on the open sea. His calculation of a diagonal route that crossed lines of both longitude and latitude was based on the computation of numerous small segments visually and conceptually integrated into one continuous whole. Since these segments would be measured in time, the route itself became conceptualized as a continuous stretch of time (see fig. 5.1).

Yoshitoki's work provided Toshiaki with methods and conceptual approaches needed for navigation. In *The New Method of Crossing the Sea* (*Tokai shinpō*), Toshiaki speculated about the actual techniques required for open sea sailing.[21] While researching the treatise, Toshiaki traveled aboard the merchant boats that plied the Japanese islands in order to gain insight

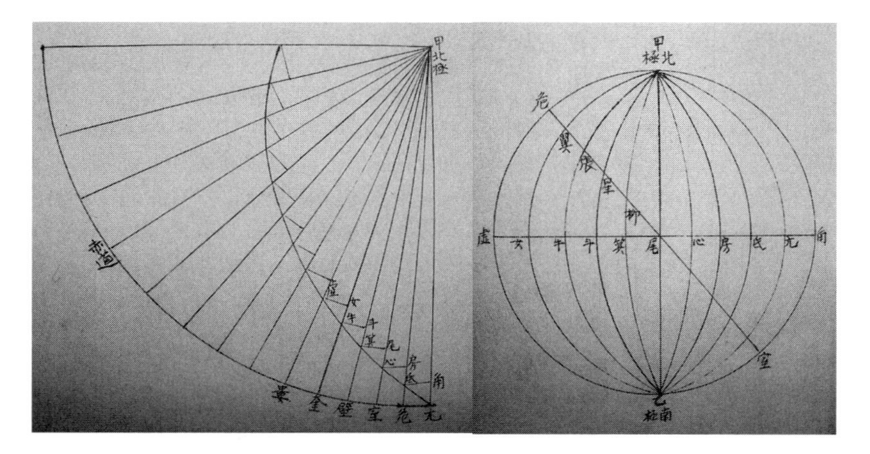

Figure 5.1. Diagrams in Takahashi Yoshitoki's *Thoughts about Seafaring Routes* (*Kaichū funamichi kō*), 1803. Japanese Academy of Science, Gakushiin Archive, Tokyo. Yoshitoki treated seafaring as a mathematical puzzle—how to chart a route on a spherical surface. Thinking abstractly, he ignored the locations of continents, currents, and political boundaries and focused exclusively on mathematics. As he showed in the diagrams above, to maintain the same angle across longitude lines one would need to sail in a curve, and not in what appears to be a straight line. In the left diagram, the top right corner is the North Pole, the bottom curve is the equator, and the lines stretching between the two are longitudes. Crossing each longitude at the same angle produces a curved line. Yoshitoki's route is a continuous aggregate of line segments, determined, in part, through the measurement of time. The practice of defining geographic location in terms of arcs and angles of time is thus translated into one continuous time-flow that wraps around the earth.

into the experience of sailing. Yet, in spite of the fact that Japan had a highly developed commercial marine infrastructure at the time, all Japanese shipping routes lay close to land. Toshiaki learned that Japanese sailors relied on familiarity with the sea floor and orientated themselves using visible landmarks on the shore.[22] These techniques could not possibly work for open sea navigation, where there were no visible landmarks and the sea bottom was imperceptible.[23]

Toshiaki could only speculate about possible solutions to the problem of orientation at sea. Extrapolating from his knowledge of astronomy and surveying methods, he proposed the following strategy for finding longitude at sea:

> On a predetermined day, starting from noon exactly, prepare the clock. Then, that night, wait until a certain big star reaches the southernmost point. Then note how much time has passed since that noon to this point and verify it with the measuring clock. This would be the time for this star. In the tables,

> estimate what should be the time of this star being in the south at this specific day at the country of origin, and register that time. Compare the measured time with the estimated one, and find the time difference between the country of origin and the time of the place where the ship is. If the measured time is less than estimated, the ship is to the west of the origin; if it is more, then the ship is to the east of the origin. Take the time difference, multiply it by 15 and you will get the difference in degrees and hence in the longitude. If the ship is to the east of the country of origin, then the longitude would be eastwards; if it is on the west, then the longitude would be to the west.[24]

Although the tables Toshiaki mentions are those of the 1795 British Nautical Almanac, the steps of preparing and watching the clock almost exactly replicated the time-measurement practices Inō Tadataka had employed in the course of his surveying missions.[25] Like Inō, Toshiaki relied on a comparison of observed durations to determine geographical location. Unlike Inō, however, Toshiaki's perception of time-space unity was not bound to the Japanese archipelago. When looking at the calculation of longitude over vast stretches of ocean (as opposed to the thin crescent of the Japanese archipelago), Toshiaki's practice of translating time into longitude acquired a slightly different meaning, suggesting that the globe was wrapped by a "belt" of time, with all the various places on the face of the earth linked together by their location on this belt. If Inō created a time/space link on a local level, Toshiaki stretched that link to global proportions.

Toshiaki's ideas about the global interconnectedness of time and space would be echoed in the writings of his many students. In 1813, mathematician Hattori Yoshitaka wrote a popular treatise on open sea navigation entitled *The Record of Safe Seafaring* (*Kaisen anjō roku*). He opened the treatise by stating that "Japan is a country surrounded by the sea, and therefore should know seafaring. But in the recent years, there is no one who employs the understanding of degrees of heaven in order to go out into the open ocean. [. . .] The Westerners know this method of navigation, but in our country there is nobody who had mastered it yet. I have studied under Master Honda and learned from him [this method]."[26] Describing the globe in an astronomical sense, Hattori's book relied on heliocentric theory. But his readers did not necessarily dwell on general cosmological questions of heliocentricity, and he himself claimed that "for navigation purposes it doesn't matter what revolves around what."[27] The method he proposed was the method of "three rates"—distance, degrees, and time. Explaining first that these three parameters were interchangeable and translatable into one another, Yoshitaka then provided a series of mathemati-

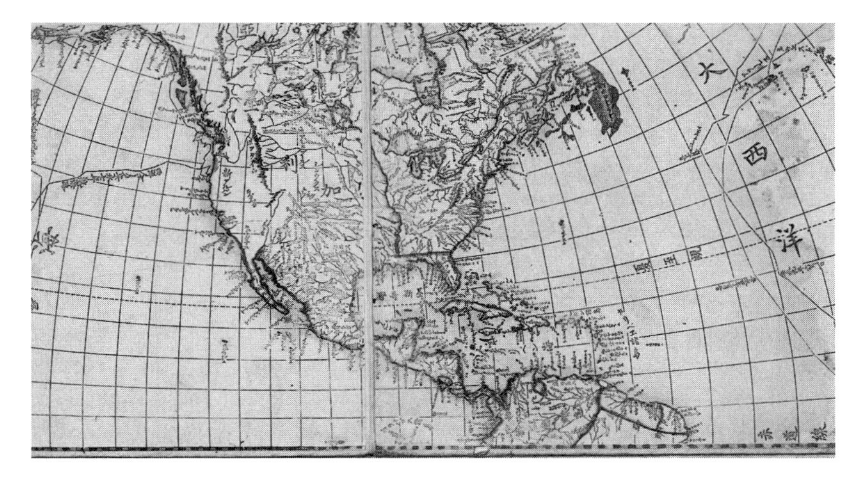

Figure 5.2. Takahashi Kageyasu's *The Newly Revised Map of All the Countries* (*Shintei bankoku zenzu*). Waseda University Kotenseki Sogo Database of Chinese and Japanese Classics. Kagesuke constructed his map using information gleaned from European books, as well as his conversations with foreigners he met in Japan. His map not only shows geographical features but also provides faraway place names, such as Boston, Toronto, and Maryland.

cal problems that challenged readers to calculate longitude, or to find the exact distance traveled by examining the available parameters. The lesson that his readers learned from following the exercises was that various geographic locations were situated on an imagined belt, defined by degrees and measured in time. By looking at world maps, such as the one composed by Takahashi Kageyasu, they could also ground their imagination of places like Nova Scotia, New York, Hudson Bay, Maryland, Georgia, and numerous others mentioned by Toshiaki, in a visually coherent global picture[28] (see fig. 5.2). Decades before the advent of European modernisms predicated on the annihilation of space and time through real-world technologies like railroads and radios, these playful Tokugawa-period calculations of hypothetical longitudes brought about the idea that time measurement connected all the different points on the globe.

Chronometers and "Chronometers"

In addition to expanding a sense of interconnectedness from the Japanese archipelago to the globe, Toshiaki's navigational method also demanded a timepiece that differed from the one used by Inō Tadataka. While Toshiaki's teacher, Yoshitoki, contended that "there is nothing that surpasses the pendulum clock for the measurement of longitudes," Toshiaki disagreed.[29]

The pendulum clock was not a good device for navigation purposes, because the rolling of the sea easily disturbed the oscillating pendulum. Instead, Toshiaki suggested using a spring-driven device aboard ships. The clock that he suggested, the "seconds-clock" that measured 86,400 seconds in one day, was the same class of European technology that provided the time measurements for British nautical almanacs and that had produced the formulation, seen in the quotation above, that one hour equals fifteen degrees.[30]

European watches were not new to Japan. Watches had been continuously imported to Japan through Dejima and modified for the local market by Japanese clock-makers, who altered them to measure time according to different systems of hours. Unlike lay consumers of these timepieces, however, astronomers were also aware of unmodified Western clocks that maintained Western conventions of time measurement. Asada Gōryū had even tried to use a Western watch for astronomical observations in 1785, but quickly gave up as neither the timepiece nor the units in which it measured time corresponded to his observational practices of the time.

Western timepieces came to be known among astronomers as *jishingi*.[31] *Jishin* was a term used to signify *invariable* astronomical hours, and *gi* is a suffix that was often used to signify an astronomical device. The term can thus be translated as "hour device" or "time machine," but its usage was more specific than the generality the term might imply. After the 1840s, the term *jishingi* was applied exclusively to marine chronometers, and professional astronomers made a point of distinguishing *jishingi* from Western-style consumer wristwatches, which were referred to as *netsuke*-clocks or sleeve-clocks (see chapter 2). But at the beginning of the nineteenth century, astronomers did not yet make this distinction.

Early nineteenth-century Japanese were not the only ones who failed to distinguish between marine chronometers and regular watches. At the time, marine chronometers were still a relative novelty. It was only in 1761 that John Harrison came up with his ultimate design of a marine chronometer, a device that could not only withstand the rolling of the sea but also maintain time continuously without constant calibration. But Harrison's device was too novel and expensive to enter common use, and even by the beginning of the nineteenth century most European ships could not afford a real marine chronometer. Instead, they used high-quality consumer watches for the purpose of timekeeping. Even though these devices were not as reliable as marine chronometers, they were considered "good enough" and hence were referred to, also, as "chronometers."

It is doubtful that Toshiaki had access to an actual marine chronometer when he suggested the use of *jishingi* for the purpose of time measurement. Adam Johann von Krusenstern, who sailed to Japan in 1804 in order to convince the Japanese government to sign a trade agreement with Russia, carried a device that he referred to simply as a "chronometer."[32] Krusenstern was not allowed to argue his cause to the shogun and instead was confined to the island of Megasaki. Frustrated by his situation, Krusenstern conducted observations. "The Japanese," he noted in his diaries "have no notion of geographical longitude or latitude." He elaborated that "in a town of not great distance from Jeddo,[33] [there are] people who inhabit the temple and [are] called *Issis*, and who possess the art of foretelling eclipses of the sun and the moon."[34] Such claims make it clear that he did not have any contact with local scholars and was basing his opinions on old European accounts. He appears to have been completely unaware of both Takahashi Yoshitoki's astronomical treatises and of Inō Tadataka's geodetic expeditions. Nor was he aware of the fact that on the very day he was disappointed by cloudy skies that concealed a total eclipse of the moon, Hazama Shigetomi was just several miles away and equally disappointed by the impossible observational conditions. Consequently, even if his "chronometer" was indeed a marine chronometer, these Japanese astronomers would not have known about it.

It is therefore likely that the first marine chronometer Japanese astronomers saw was the one brought by Philipp Franz von Siebold. Siebold had come well equipped to Japan in 1823. On his trip to Edo, "besides a barometer and Torricellian glass tube for altitude measurement, hygrometer and thermometer," Siebold carried "a chronometer of Hatton & Harri[s], London, a sextant also made in London, supplied with a nonius from which one could read 15 seconds; artistic horizon and compasses, etc., as well as electric and galvanic apparatuses, several compound microscopes, and finally, a small fortepiano."[35] Hatton & Harris was a shop that produced eight-day marine chronometers, which suggests that Siebold's device was not merely a "good-enough" wristwatch but the real thing: a marine chronometer.

Siebold demonstrated his chronometer to Japanese astronomers, using it at least five times during his trip to Edo.[36] He made some longitude measurements for his own records, and several times conducted public measurements to entertain his varied royal hosts. It is clear that Japanese astronomers were present during several of these demonstrations. Some, like Takahashi Kageyasu, got to discuss scientific observation with Siebold in person, and even developed personal relationships with him. When Siebold was preparing for his departure from Japan in 1828, he promised

Kageyasu that he would send him a real marine chronometer from abroad or bring one in person on a planned return trip. However, after the unfortunate events surrounding the discovery of Inō's maps in Siebold's possession and Kageyasu's subsequent death, this plan was never realized.[37]

The arrival of the marine chronometer substantially changed Japanese astronomers' evaluation of pendulum clocks. As had been the case in Europe and America, marine chronometers remained largely inaccessible in Japan long after their introduction. Also, like many Western pilots, Japanese astronomers had to compromise by using "good-enough chronometers"—namely, high-quality Western wristwatches. But even "good-enough chronometers" seemed to offer solutions to technical problems, which had been tolerated until then.

The person who most actively promoted the use of Western watches in astronomy, and who perhaps played the greatest role in the reformulation of the notion of time as singular, global, and Western, was Shibukawa Kagesuke.[38] The youngest of Yoshitoki's sons, Kagesuke was adopted into the Shibukawa family of astronomers after Yoshitoki's death. Following the adoption he officially became a descendant of the seventeenth-century astronomer Shibukawa Shunkai. Kagesuke made it his life's project to continue the work begun by his father and elder brother, editing and revising their works. Kagesuke's "revisions," however, amounted to major changes, both in volume and content. His "addendum" to his father's *Thoughts about Seafaring Routes* far exceeded the page count of Yoshitoki's own ruminations, and evolved into a multivolume corpus dealing not only with the calculation of seafaring routes but with a variety of navigational practices. Kagesuke's advanced mathematical skills and knowledge of Western astronomy allowed him to correct his father's calendrical algorithm and to devise a new system that would serve as a basis for the final astronomical reform of the Edo period—the 1844 Tenpō reform.

In similar fashion, Kagesuke's attitude toward timepieces and time measurement was both a continuation and a major revision of his father's and brother's practices. In his *Account of Divine Constitution* (*Reigen kōbo*), Kagesuke stated that "already Hazama Shigetomi and Takahashi Yoshitoki noted that when candles are extensively used in the proximity of the [pendulum clock], or when many people gather around, the sequence of swings is lost. Or, the same holds when the weather is extremely cold."[39] The older generation—Takahashi Yoshitoki, Hazama Shigetomi, and Inō Tadataka—had not treated this problem with the pendulum clock as a critical shortcoming, and instead had formulated what was for them the reasonable solution of calculating the duration of a single oscillation by dividing the length of a day by the total number of oscillations in a particular daily cycle, which they deemed to be "pretty much precise."[40] But for Shibukawa Kagesuke even a variation of just three or four oscillations from day to day constituted a major problem. The fact that the pendulum was essentially useless aboard a ship made this device utterly flawed in Kagesuke's eyes. Consequently, he decided that the "precise construction" of the Western "Three-indexed Time Machine" made it much more suitable for astronomical observation than the pendulum clock.[41]

As the idea that Western watches were more suitable for astronomical observation took hold, astronomers came to measure time exclusively using their mechanisms. As a result, Japanese astronomical timekeeping would go through another major revolution—by the 1840s and '50s astronomical calculations no longer relied on the astronomical *koku* or on the oscillating pendulums. Instead, all calculations came to be based on the Western system of 24 hours, 60 minutes, and 60 seconds, or, 86,400 seconds per day.

Time as a Mathematical Average

Kagesuke's preference for Western clocks was not motivated by practical reasons alone. After all, Western watches were not without their flaws. In 1842, for example, a Confucian scholar and clock collector, Satō Issai, noted that he had had to lend one watch to Kagesuke because the latter's had broken, and Kagesuke had urgently needed one to observe a lunar eclipse.[42] And since most of the Western timepieces Kagesuke used were not marine chronometers, but watches, even the ones that did not malfunction could be affected by the environment. At the same time, the fluctuations in the number of daily oscillations of the pendulum clock were known almost immediately after Hazama Shigetomi invented the device, yet they were not perceived as a serious problem before Kagesuke's time. Rather, what shifted Kagesuke's evaluation of timepieces was his perception of time measurement. Unlike his predecessors, Kagesuke saw the time of the Western watch, and certainly of a marine chronometer, as a qualitatively different kind of time. For him, Western watches did not just measure temporal units—they offered a mechanical representation of the physical reality of the universe.

Kagesuke perceived Western watches as qualitatively different because they kept mean time. The idea of mean time was itself inherently related to the development of marine chronometers. Over the course of Christiaan Huygens' repeated yet unsuccessful attempts to design a timepiece for navigational use, he had begun contemplating a phenomenon long known yet still neglected in his time. This phenomenon was the slight variation in the length of the apparent day—the time measured between two consecutive observations of local noon. This previously unproblematic astronomical fluctuation had become crucial to the calculation of longitude. After all, aboard a ship navigators could determine only the apparent noon, the one that changed from day to day; but the chronometers they carried measured days of uniform length and were set to a local noon at a place of origin on the date of departure. To solve this problem, Huygens added all the varying values of apparent days and divided them by 365, thus producing a mathematical average to match what we consider to be exactly 86,400 seconds. According to this equation—known today as the equation of time—the variation follows a fluctuation path in which on only four specific dates the day lasts exactly 86,400 seconds.[43] The average produced by the equation of time—known as mean solar day—became the standard for marine chronometers. Roughly one hundred years later, England began publishing nautical almanacs, which took the needs of sailors into account by providing data in apparent solar time, yet included instructions on how to calculate the difference between the observed apparent time and the mean time kept by the ship's chronometer on any given day.

Like their European colleagues, Japanese astronomers were aware of fluctuations in the length of the apparent solar day, but before their interest in navigation, they had not considered it crucial to their observational practices. Timepieces, regardless of the units in which they measured time, were calibrated according to local noon. Consequently, if two astronomers made simultaneous observations, they calibrated their clocks on the same day and thus relied on the same standard. And when astronomers made time measurements of a regular event, such as a solstice, they would do it on the same day of the year, meaning that the apparent length of the day on which they calibrated their clocks was consistent.

Nevertheless, while engaging with the 1795 British Nautical Almanac, Yoshitoki brought the phenomenon of the fluctuation in the length of the apparent day out of its relative obscurity. In his notebook from 1801, entitled *Study of the English Almanac* (*Angeria reki kō*), he attempted to calculate the longitude of important Japanese locations such as Edo, Kyoto, and Nagasaki, comparing the times of his observations with the foreign data. But he was perplexed by the fact that the data in the British nautical almanac differed slightly from that of other books. Even when he accounted for differences in longitudinal references (the British almanac counted longitude from the observatory in Greenwich, but the Dutch placed the prime meridian on Tenerife), the values Yoshitoki calculated using different sources were always slightly different from each other.[44] He noticed, however, that one of the largest discrepancies occurred when he used the timings of celestial positions that occurred during May:

> When I calculated the longitudinal difference between Eastern capital (Edo) and London according to Lunar Distance, it was 9:19:20, which differs from the above 9:25 by as much as 5 min 40 seconds. In the English almanac the times of the five stars are given in the apparent time, while the calculations I made were in the mean time. If we take the times of the Western fifth month, which is [the end of] our third and [the beginning of our] fourth month, the time difference is 3 minutes 59 seconds.[45]

The pure chance of choosing to work with data from mid-May highlighted the phenomenon of fluctuation in the length of the apparent day. Mid-May is indeed one of the peaks in the fluctuation, when the length of the apparent day diverges from that of the mean solar day by almost four minutes. It was this large discrepancy that made Yoshitoki consider the difference between mean and apparent time. Unfortunately, accounting for this difference still did not solve his problem calculating longitudes, but his engagement with mean time established it as a potentially important factor in astronomical calculations.

After Yoshitoki's death in 1804, the discussion of mean time went dormant until Kagesuke resurrected it several decades later. In 1828, the year of the Siebold incident that would cost the life of his elder brother Kageyasu, Kagesuke acquired another British nautical almanac. Kagesuke saw his writing as an expansion and extension of his father's work, and his initial approach to mean time closely followed Yoshitoki's by providing the time difference between Greenwich and Edo.[46] Unlike Yoshitoki, however, Kagesuke had no doubts about the role of the difference between the mean and the apparent time in his calculations. The tables were given in apparent time, and hence the observation times were dependent on the time of the year. Longitudinal difference, on the other hand, is constant and in order to calculate it using observed times one had to incorporate the equation of time (calculated by Huygens and provided in the almanacs) into the calculation.

But there was an even more fundamental difference between Yoshitoki's and Kagesuke's approaches to mean time. For Yoshitoki, mean time was just one of several kinds of time one could work with, and it was up to him to decide which situation called for which kind of time. When he used the

British nautical almanac to advance his astronomical calculations, mean time had served as just another parameter for which he needed to account. But Japanese interest in navigation evolved over the course of the next two decades, such that when Kagesuke read the almanac of 1828, he came to see mean time as relevant to both seafaring and the specific timepieces that displayed it. For him, mean time was not merely an alternative way to *divide* the day, but also an alternative way to *define* the day.

For Kagesuke, the Western watch embodied the concept of mean time. In a short treatise, *Questions and Answers about Chronometers* (*Jishingi montō*), he summarized what he believed was the function of the Western watch[47]: "The hours of the Western miniature clock," he wrote, "are determined by one revolution of the sun on the ecliptic to the same position it was before [. . .] Calculating the [daily] average it comes to twenty four hours a day, [. .] which is called 'mean time' [. . .] But the hours used in Japanese calendars or announced by bells are all according to the 'common time,'"[48] which Kagesuke explained as "the sun's rotation from one southernmost position to the same position the following day." In order to make this point clearer, Kagesuke offered an example of the difference between mean time and what he called "common time" as well as the calculations associated with it:

> So, [what do you do] if you want to know what the watch should read when you have measured the "common" 12 o'clock as the true noon with the sun at its southernmost position, on the seventh of the first month, this year? First of all, looking at the time equation tables you see that that day's difference is 14 minutes, 32 seconds (I explain the time equation later).[49] Consequently, adding the universal difference to the normal 12 o'clock (Usually, the time equation is used to distinguish common time from mean time, but here we take common time as a reference point and infer mean time from it. Consequently, instead of the usual subtracting, we add the value), we get 12 o'clock, 14 minutes and 32 seconds. This is the mean time as opposed to the common time. Therefore, when it is 12:14:32 on the watch on the seventh of the first month of this year, you know that it is the ninth hour [namely the noon, according to Japanese reckoning].[50]

For Kagesuke, the difference between the time shown by a Western watch and the time commonly used in Japan was explained by the fact that the latter indicated "common time," while a Western watch showed "mean time"—the average period of the sun's motion relative to the earth.[51]

Kagesuke's argument is all the more fascinating because the connection between mean time and Western clocks is not in fact inherent. Mean time is a mathematical average, and thus is unaffected by the specific temporal units in which time is measured. Units of measurement, of course, can be adjusted to fit any precalculated value necessary. (In fact, the duration of one second, which was calculated as an astronomical average in early nineteenth-century Europe and which we still use today, no longer matches the astronomical average due to changes in the length of the solar year.) Hypothetically, the Japanese astronomical system of decimal units could have been adjusted to measure mean time in a manner similar to Huygens' of the previous century. On the other hand, most Western timepieces of the time did not actually maintain mean time at all. During the first half of the nineteenth century many Europeans—and certainly the majority of Americans—calibrated their timepieces using local noon.[52] Consequently, even if these Western clocks measured minutes and seconds calibrated to reflect the mathematical mean time, except for four specific days each year, when the length of the apparent day coincides with this average, Europeans' and Americans' clocks measured either less or more than 1440 minutes in a day.

But the confusion created by the term "chronometer," in the way it was used both in European languages and in Japanese, shaped the illusion for Kagesuke and others that all Western watches kept mean time. According to this belief, the time kept by Western watches, and only by Western watches, was calculated mathematically and based on astronomical reality. It was independent of human activity and not even necessarily tied to the observable local noon (which required an observer). As such, for Kagesuke, the time indicated by the Western watch became *the* true time.

Time Begins at the Prime Meridian

When open-water navigation became a reality in the final years of the Tokugawa period, the new market for navigational manuals was shaped by Kagesuke's writings. Fujino Masahiro's 1861 *Explanation of Navigation Method* (*Kōkai yōhō setsuyaku*) was purportedly a translation of Jacob Swart's *Zeevaartkunde*. But, similar to Kagesuke's incorporation of texts he learned from his father into his "translations" of Western books, Masahiro frequently quotes Kagesuke almost verbatim. And in his preface, Masahiro admits that "in order to make sense in Japanese" he had to change Swart's text. Masahiro opened the treatise by stating something that by his time had become obvious: "If you want to go to the foreign countries, you should first learn the shape of the earth and its geography. These detailed explanations are naturally found in geography books that have all kind of maps—you should look at all of them."[53]

If knowing the shape of the earth and the relative positions of all the countries was not exactly necessary to all specific practices aboard a ship, such knowledge *was* necessary to strengthen the global imagination and visualize the binding flow of time, so essential to the ability to calculate one's position at sea. Reading the Dutch text through the lens of what was already known to him, Masahiro described the distinction between three types of time: sidereal, solar, and mean. "Sidereal time is the time that it takes a certain star to come back to the same position after it has passed the meridian. Solar is from one crossing of the meridian by the sun to another. However, the time that it takes the stars to complete one round is not the same as for the sun and it is faster by approximately one degree [. . . .] But there is also difference between the solar and the mean time. The rate at which the sun moves westwards every day is not the same."[54]

In the same year that Masahiro published *Explanation of Navigation Method,* Tokura Sukeyuki published *Insights into Navigation* (*Kōkai hitsudoku*), another purported translation of Western navigation literature.[55] He, too, repeated the known facts: "The sun makes its way from East to West and in 24 hours (our 12 hours) makes one round of 360 degrees. Consequently 15 degrees make an hour, 1 degree makes 4 minutes, 0.1 degree is 4 seconds. Therefore, multiplying and dividing by 15 and 4 you can change longitude into time or time into longitude." Like his predecessors, Sukeyuki stressed the crucial importance of distinguishing between mean and apparent time: "If you set your timekeeper according to noon, there might be a difference between it and the mean time, in the place where the ship is."[56] In order to train his readers to deal with this difference, Sukeyuki provided a series of mathematical problems aimed at calculating the difference between mean and apparent times using the equation of time (found in the nautical almanacs).

By 1861 it had become clear that not all Western watches were chronometers, and thus not all kept mean time. Thus, Sukeyuki made a point to distinguish between the regular Western watch that kept the apparent time (since it was still calibrated according to the local noon) and the marine chronometer that kept true mean time—although regular watches were a good-enough substitution when there was no better alternative available, their relative inadequacy affected actual navigational practices.[57] But a specialist's distinction between regular watches and marine chronometers did not compromise the popular association of the Western temporal system with mathematical—and thus universal—reality. Western time, in 1860s, stood for astronomically and mathematically "true" time, which also happened to serve the new practical objective—sailing abroad.

When longitude calculations ceased to be hypothetical, another problem emerged—where on the globe did this mathematical and universal time originate? As Masahiro puts it: "For latitudes, the equator serves as a reference point, and all the latitudes are calculated from it. For longitudes, there is no such a place." "Therefore," he continued:

> Each country decides where [time] starts. England sets its prime meridian to the place called "Greenwich" where they have an observatory. But France sets its prime meridian to Paris, in Spain it is a place called Cadiz and the Dutch take the Pico of Tenerife. Recently, both America and Russia decided on Greenwich as their prime meridian.[58]

The obvious conclusion from the fact that every country had independently decided from where to measure longitude was that Japan was entitled to do so also, as it had with its determination of terrestrial longitude on Inō's maps. "However," Fujino Masahiro continued:

> Although our country obviously should measure its degrees relative to Kyoto, our countrymen have not yet sailed abroad and therefore have not measured the longitudes and latitudes of various places. Meanwhile, we use the longitudes and latitudes determined in England, and only after taking this as our reference point, calculate the longitude of Kyoto. According to English calculation, Kyoto is at 135° 40′ longitude.[59]

Under these circumstances, as much as Japanese scholars would have loved to determine their own prime meridian, practical considerations made it impossible. There was nothing specific to the Western division of the day into 86,400 seconds or about the location of the Greenwich observatory that made either better suited to the measurement of longitude, but by the mid-nineteenth century, Greenwich-based time had become pervasive. The comparison tables brought to Japan all pointed to Greenwich time and the marine chronometers brought to Japan aboard the foreign ships all measured time according to Greenwich. The abstract time introduced with visions of navigation and chronometers was not only global—its prime meridian lay far away from Japan.

Coincidentally, however, the Greenwich-based global time reinforced the Japanese idea of national time. In many instances, the division of the globe into time zones created a situation where the "center" of a time zone was far away from a national capital, and sometimes even in a different country altogether, resulting in confusion, if not outright opposition.[60] Kyoto, however, happened to fall almost exactly in the center of the time zone of Japan, calculated to be nine hours ahead of Greenwhich local time. Kyoto was not, at that point, the capital of Japan. Nor was it the center of bureaucratic or marine activity. However, as we have seen earlier, there was a longstanding tradition of treating Kyoto as the center of Japanese time. Thus, even though the Greenwich-based global system located the center of time far away from Japan, the center of the Japanese *time zone* was still aligned with the conventional perception of Kyoto-based national time.

* * *

Preference for Western time thus emerged in the middle of a hypothetical ocean, with no landmarks in sight to orient the ship. It grew out of navigational timekeeping practices during a period when no Japanese were actually permitted to travel on the open sea. The processes of the globalization and Westernization of time did not happen only in response to already existing practices, but rather evolved out of imagined and projected demands. It is only by looking at our protagonists' field of knowledge, and exploring how they extrapolated from them, that we can appreciate how these individuals made their desired futures into reality.

By the end of the eighteenth century, Western temporal conventions were well understood and even used by Japanese astronomers in their calculations. Yet Western time was just one of the possible kinds of time they juggled, and it was only applied when and where it appeared to be the optimal option. The transformation in the status of Western time began when it came to be associated with navigation and was superimposed on a global imagination of open sea travel, commercial seafaring routes, and distant countries. Time was a known entity in the vast ocean of the unknown, and it provided a conceptual anchor to which routes and places could be tied. Due to the structure of Western timepieces, Western time was the only kind of time that could be measured at sea, which meant that global time was Western time.

Timepieces also played a crucial role in the development of a new characterization of Western time as distinct from and truer than other competing temporal systems. Whereas in the beginning of the nineteenth century, mean time was just a calculational option, its indispensable role in navigation gradually prioritized it relative to other kinds of time available to the Japanese. And the optimistic if inaccurate association of all Western timepieces with this mean time brought about the notion that Western temporal conventions stood for universal mathematical and astronomical realities.

Even though the global and astronomical dimensions of mean time made it more important than the other kinds of time, they also formed the crux of a problem: if time is independent of human activity or observation, how do we decide where it starts? The answer to this conceptual problem, for Japanese astronomers and aspiring navigators, fell short of the epistemological significance suggested by the question. Time started where people with the most data said it did. Tying oneself to Western time meant tying oneself to Western longitudes. And in the battle of nautical almanacs in Japan, the British nautical almanacs, filled as they were with data measured according to the Greenwich prime meridian, prevailed.

There was no single Westerner who propagated this notion of time in Japan, and there was no Western text that explicitly instructed Japanese how to understand Western time. Unlike other non-European countries, Japan was not forced to accept Western timekeeping, since at that time, Europeans had neither military nor political nor cultural dominance over Japan.[61] Nor can we say that mechanical clocks alone revolutionized Japanese time, since the ways they were "read" in Japan were initially rooted in local practices and associations. Detached from cultural context and associated practices, mechanical clocks could not change the whole array of associations with timekeeping prevalent in Japan. However, extrapolating from the bits of astronomical practices with European roots, Japanese astronomers and aspiring navigators could make conclusion about timekeeping that were reminiscent of their European counterparts'. They were by no means identical, but they were similar enough to make global communication about time possible.

Marine timekeeping was on the minds of Japanese astronomers long before they could actually try it out during open-sea navigation. But even though measuring time with Western watches or chronometers onboard ships started as a thought experiment, it evoked a very real transformation in perceptions of timekeeping practices. The hypothetical was extrapolated from existing practices and existing associations with timekeeping. Once this hypothetical idea took root in the discourse of early nineteenth-century Japanese scholars, it became transformative in itself, effecting the ways people perceived the measurement of time.

Times had changed. Moreover, it is even possible to say that in a certain sense "Time" had emerged—that is, the modern, singular notion of time. There was actually nothing inherently "modern" about Western mean time—it represented neither modes of production nor a disassociation from the supernatural, nor was it more distant from "nature" than previous notions of time in Japan. Neither was it inherently astronomical, as any units could be adjusted to represent the mathematical average of the time equation. Mathematical, astronomical, and independent of human activity, *this* time was not modern in itself, it was just the kind of time we happened to use during the historical period we decided to call "modernity." And it was this notion—based on the primacy of astronomical practices—that became associated in Japan with Western time.

SIX

Time Measurement on the Ground in Kaga Domain

In the previous chapters we have seen how, through their computational and observational practices, astronomers, geographers and aspiring navigators came to see time measurement in terms of arcs, angles, and diagrams. Time came to be seen as corresponding to the motion of the stars or the motion of hypothetical ships in the ocean. It was inscribed in maps and navigational charts. This evolving notion of time measurement, however, was not limited to a select group of high-ranking specialists. Let us go back to the beginning of the nineteenth century and explore how this emerging set of associations with time measurement spread through networks of scholars, intellectuals, and educated officials. Specifically, we will focus on one individual, Endō Takanori, whose work exemplifies the reach of astronomers' changing conceptualization of time measurement.[1] Both his professional endeavors as well as his idiosyncratic personal interests reflect the early nineteenth-century marriage of celestial motion, time measurement, geographic locality, and linear perspective.

Takanori was a high-ranking official of Kaga domain. The territory of former Kaga domain roughly overlaps with the southern half of present-day Ishikawa prefecture, lying on the inner curve of the main Japanese island, on the side of the Japanese sea. Surrounding the *daimyō*'s castle, the city of Kanazawa was the capital of Kaga, and it remains the capital of Ishikawa prefecture today. During the Edo period, the numerous domains of the Japanese archipelago all had their own unique characteristics, and the Kaga domain, too, occupied a rather special role. It was close enough to Kyoto and Osaka to be involved in the latest cultural developments, yet far enough away to develop as an independent cultural center on its own. It was also fertile and rich—the richest of all the domains. Kaga's resources allowed its rulers to invest in material, social, and cultural infrastructure, build highways and water-control systems, develop poverty relief programs, accumulate libraries full of books, and invite famous scholars to reside and work there. At the beginning of the nineteenth century, it was a home to famous thinkers, astronomers, and doctors practicing Western medicine and was an essential destination for traveling scholars of all specializations.[2]

Takanori enjoyed the material and intellectual resources provided by his birthplace and his high rank. From the various visiting and local scholars he learned astronomical and surveying practices, the operation of a variety of timekeeping devices, philosophy, economics, medicine, drawing, and other skills. As a result of his education, he adopted observational techniques that tied time measurement to the movements of celestial bodies. He relied on calculation methods that expressed this association in triangle diagrams and connected both to the specific geography of Kaga. By doing so, Takanori developed an approach to time measurement that represented time as rooted in geographical and astronomical reality.

Endō Takanori's Astronomical Education

In 1799, the fifteen-year-old Takanori officially began his studies with astronomer Nishimura Tachū.[3] Tachū was a Kaga native who left the domain to study astronomy with Kyoto-based Nishimura Tōsato. Tōsato was the author of *Illustrations of Various Measurement Devices*, which described the timekeepers of various Japanese astronomers, and was one of the most fervent critics of the Hōreki calendar.[4] Tōsato's teachings formed the basis of Tachū's approach to astronomy, particularly the latter's life-long interest in measurement instruments. Just four years after Tachū began his studies, Tōsato passed away and Tachū moved to Osaka to study with Asada Gōryū but retained the last name "Nishimura" to honor his late teacher. Under Asada Gōryū, Tachū learned calculation and observation practices described in Jesuit Chinese astronomical treatises and became close friends with Takahashi Yoshitoki and Hazama Shigetomi. By the end of the 1790s Tachū had returned to Kaga to assume the duties of a domain astronomer, yet he remained in frequent communication with Asada Gōryū's other students, updating them about his observations in Kaga and consulting with them about the problems he encountered. In 1799 Tachū was officially appointed as a tutor to young Endō Takanori, whose education was geared toward his future career as a minister of infrastructure of Kaga domain.[5]

Already during the early years of his studies, Takanori struggled with the unique timekeeping system of Kaga domain. One of the first assignments Tachū gave Takanori was to explore the time-measurement methods that existed in Kaga. In the course of his investigation, Takanori compared the measurements made with a sundial from 1758 with another set of measurements that he himself made with a sundial designed on the basis of astronomical calculations as taught by Tachū. He found that the time measurement by the old sundial, called Time Known through Sun's Shadows (*Kichiji*), differed from that measured by the later device, called Sun's Shadows Timepiece (*Kijiki*), with the older one giving time that was "late in the morning and early in the evening."[6] Attempting to explain the discrepancies between the two, Takanori pointed out that the old sundial was built to measure time according to a very peculiar temporal convention of Kaga domain. Being economically and culturally independent, Kaga domain also had a temporal system that differed slightly from the rest of the country. As we learn from Takanori's description in the *Record of Kanazawa's Time Bells* (*Kanazawa jishō ki*), the people of Kaga domain had not twelve but thirteen hours in one day.[7] They still used the twelve animal signs and the two consecutive series from nine to four in order to count the hours; and they, too, divided the day into daytime hours and nighttime hours, so that all the hours were of variable length. The peculiar feature of the Kaga calendar was the inclusion of an "addition" (*amari*) before dawn and another after dusk.[8] Each of these "additions" lasted approximately half of a Japanese "hour," and two of them—one at dawn and another one at dusk, added up to another full hour. Looking at the measurements made with the two sundials—one based on this convention and another on the astronomical calculations of Nishimura Tachū—Takanori had no difficulty spotting the difference. This was not a simple question of inconsistency. Under Tachū's tutelage the young Takanori had already embraced notions of time defined by the movement of celestial bodies. Thus, from a young age Takanori saw measuring of time as representative of astronomical reality, establishing the criteria for what was "correct" or "incorrect" about it.

This perception of time measurement as significant representation of physical reality was further reinforced by Takanori's early encounters with land surveying. Learning the practices involved in measuring land, Takanori conceptually tied time measurement not only to astronomical observation but also to geographic locality. Takanori first became aware of land surveying in 1799 when a major fire broke out in the castle of Kaga's capital, Kanazawa, destroying the existing maps of the domain. At the time, a local mathematician, Ishiguro Nobuyoshi, was ordered to undertake a surveying project to create a new set of maps.[9] Japanese mathematics had always had a very strong practical aspect, and already in the seventeenth century the field of mathematics had developed a close relationship to surveying. By the beginning of the nineteenth century, utilitarian philosophy permeated Japanese scholarly circles and was especially popular among scholars who held government posts. For Nobuyoshi, surveying was not only a skill but a manifestation of the moral dimension of mathematical practice. For him, mathematics contributed to society and to the exploration of the world. Through Nobuyoshi, Takanori was exposed to scientific enterprise that would become both personal interests and professional duties as a Kaga official who oversaw domain infrastructures.

In 1803 Takanori became aware of the interconnectedness of surveying and time measurement when Inō Tadataka passed through Kaga on one of his surveying expeditions. Inō and his group met with Tachū and Ishiguro and demonstrated their latest astronomy-based surveying techniques, as well as their famous precision instruments—the theodolites and the pendulum clock. As Inō and Tachū both grew out of Asada Gōryū's school of astronomy, they already enaged in similar astronomical practices and shared similar cosmological outlooks. Thus, Kaga scholars eagerly adopted the use of Inō's instruments and embraced the intrinsic association between time, geographic locality, and the position of celestial bodies. As their student, young Takanori was educated to respect this association as a basic pillar of astronomical practice.

This association was further reinforced in Takanori's education through the teachings of Honda Toshiaki, who had been born in Kaga.[10] As mentioned in the previous chapter, the road to effective governing, for Toshiaki, started with the sciences. The word that Toshiaki used for "science" was *kyūrigaku*—a term freighted with hundreds of years of philosophical debates about the nature of human exploration of the universe in both China and Japan.[11] But Toshiaki altered the definition according to his own understanding of Western science, claiming that *kyūrigaku* was nothing but "the study of heavens and earth," rooted in "mathematical relations, [astronomical] prediction, and [observational] measurements."[12] Toshiaki also praised Western modes of depiction, which he deemed to be precise, "like the real thing," and useful.[13]

Although Takanori did not officially study under Honda Toshiaki, the latter nevertheless had a profound effect on Takanori's thought. Toshiaki was based mainly in Edo, but he visited his native domain on several occasions, during which Takanori had the chance to meet him. More importantly, Takanori indirectly absorbed Toshiaki's teachings through one of his closest friends and collaborators, Misumi Fūzō, who traveled to Edo in 1811 to become an official student of Toshiaki.[14] There are many ways in which Toshiaki's vast scholarship influenced Fūzō, Takanori, and others—

from Toshiaki's economic concerns to his views of science as a source of prosperity to utopian visions of the West. For the purpose of our story, however, it is important to note Toshiaki's geographic determinism—the belief that geographic locality has a qualitative influence on the character of the people residing in it.[15]

Establishing Kaga's "True" Time

As a government official of Kaga domain, Takanori sought to realize his scientific ideals by reforming Kaga's timekeeping practices. From 1811 to 1813 he worked on establishing a set of rules for striking the Kanazawa castle bell in order to make the announcement of time more precise. This work, however, would benefit only those who could hear the castle bell, namely the citizens of the capital and its surroundings. But what was a traveler—or even a farmer—to do when he was too far from Kanazawa castle to hear the bell? To address this, Takanori devised a portable sundial that could be used everywhere and called it Sun's Shadows Measuring Tile (*Sokkibai*)[16] (fig. 6.1). There was nothing new about using a portable paper sundial, of course, but the structure of Takanori's creation was different from most of the Japanese sundials of this period. Other sundials used a neat yet approximate diagram based on the official calendar of the central government. But having learned early in his life to be suspicious of the precision of customary practices, Takanori rebelled against conventional approximations and decided to create a diagram of his own. He conducted observations of the movement of the sun's shadows throughout the different seasons and came up with a graph that was specific to the local time in Kaga domain. By doing so, Takanori not only transformed the sundial into a precision timepiece, but also, by mass-producing it, ensured the maintenance of time that was specific to Kaga.

Takanori's perception of what made time measurement "Kaga-specific" differed greatly from the conventional wisdom. For Takanori, "Kaga time" was not only defined by local tradition but also determined by the relationship between Kaga's geographic locale and celestial motion. This perception of time was fully supported by the *daimyō* of Kaga domain, Maeda Norinaga.[17] Lord Maeda shared many of Takanori's convictions and ordered Takanori to prepare a Kaga-specific calendrical reform that would eliminate the vulgar "addition" and base the management of the domain's time on local astronomical observations.

In preparation for the reform, Takanori had to conduct additional observations, for which he needed additional instruments. In 1820 he

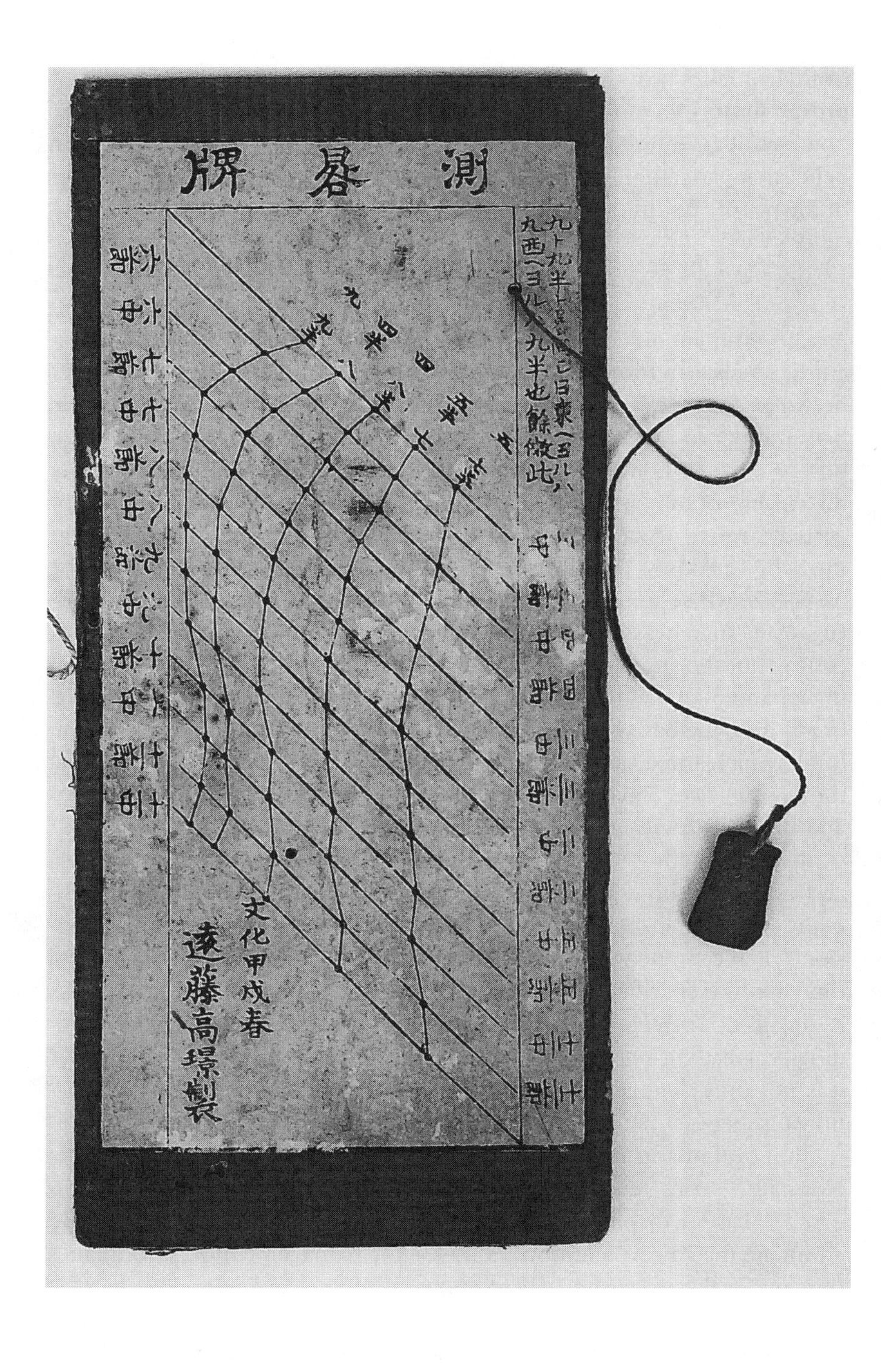
測晷牌
文化甲戌春
遠藤高璟製

designed another sundial, similarly titled Sun Shadows Measuring Plate (*Sokkiban*).[18] This was a type of equatorial sundial that was somewhat similar to the device used by Shibukawa Shunkai, as Takanori himself noted. It consisted of an inclined round plate that was set on a tripod and aligned to face north. The gnomon was perpendicular to the plate and silk strings ran between the gnomon's ends and the southernmost edge of the plate. As in any other sundial, the gnomon could be used to tell the daytime hours. The major innovation of this *Sokkiban*, however, entailed the triangle formed by the silken strings stretched from it. According to Takanori's instructions, one would note the moment when the shadow from the two strings created one single line in order to know the exact moment of the local noon.[19] Conversely, on cloudy days, one could look directly toward the sun from beneath the strings and note the moment when the sun was exactly above the meridian. In fact, this triangle of strings uses the same structure as the Meridian Device (*shigosengi*) used to create the Kansei calendar—a structure that was designed to determine the moment of passage of the sun or another celestial body across the meridian.

Ever faithful to his conviction that time measurement was inherently connected to astronomical observations made at specific geographical localities, Takanori sought to establish a calendar that would represent a kind of time that was culturally, climatically, and geographically true to Kaga domain. In order to do so, Takanori needed to maintain what was considered to be the highest standards of astronomical measurement of his day and at the same time differentiate his measurements from those that had been calibrated to the latitude of Kyoto and had provided the basis for the Kansei reform. Though Kaga lies only 1.555 degrees north of Kyoto, even such a slight difference would have amounted to an unforgivable approximation to Takanori. What Takanori sought was time measurement that was true in absolute terms, a goal reflected in the name of his new system, A True Time System (*Seijikoku*).[20]

There were several characteristics that made Takanori's temporal system

Figure 6.1. Endō Takanori's Sun's Shadows Measuring Tile (*Sokkibai*).
Ishikawa Prefecture Museum of History, Kanazawa.
The diagonal straight lines on the device represent the annual passage of months. The bent lines on the graph indicate the daytime hours. To Takanori, the unevenness of the hour lines testified to the precision of his calculations—suggesting no approximation or beautification of the graph. In order to operate the device, one lays it flat and extends the string vertically to create a gnomon, ensuring that the weight hangs directly above the string's origin. The intersection of the shadow cast by the string with the appropriate diagonal line shows the hour.
The inscription reads "Made by Endō Takanori, Spring 1814."

"true" to Kaga. First, Takanori decided to incorporate the cultural convention of an additional hour into the standard count of the twelve hours, albeit in a way that accorded with his astronomical standards. Similar to the Kansei calendar, Takanori's system specified the times of dawn and dusk not in time units but rather in degrees of the sun's motion. However, unlike the Kansei calendar, which set the moment of sunrise as the sun's position 7.3611 degrees below the horizon, Takanori's calendar established sunrise as the sun's position at 13.6974 degrees below the horizon in order to include the "addition" in the normal hour count. Consequently, the day in Kaga domain always started when it was still completely dark.

Second, Takanori sought time to be "true" climatically. Similar to the customary system in use throughout Japan, the new calendar in Kaga domain did not change the actual lengths of the day and night hours on a daily basis, but rather according to the yearly mini-seasons (*sekki*). Yet unlike the rest of Japan, which used twenty-four mini-seasons, Kaga domain under Takanori's calendar employed forty-eight mini-seasons, each of which reflected seasonal changes in the lengths of night and day.[21]

As the previous two examples reveal, Takanori's concept of true time also necessitated a higher degree of precision. When determining the beginning and end of each hour Takanori strove to capture the precise altitude of the sun, using divisions of time that were as minute as possible. Thus, for example, his calculation of the beginning of the eighth Japanese hour of the day during the equinox was set at 6.74705 *koku*, or 5829.4512 modern-day seconds after the local noon.

In order to arrive at such precision Takanori designed a special pendulum clock, unsurprisingly naming it The Board of True Time (*Shōjiban*).[22] By then, pendulum clocks such as Hazama Shigetomi's Suspended Swinging Disk were in common use by astronomers. However, astronomical pendulum clocks did not have to be consistent with any system—they only had to maintain consistency within the measured arc of celestial motion. Takanori sought to bring astronomical precision to everyday life, meaning that his clock would have to register the system of variable hours—to count time in decimal units that could be fit into 100 *koku*. Existing astronomical pendulum clocks counted roughly 60,000 oscillations per day—an unprecedented precision at that time in Japan. But "roughly 60,000" did not fit into the decimal divisions that Takanori needed. Since he had no means to raise the precision of the clock to measure 100,000 units per day, he had to compromise, reducing the precision to 10,000 units per day—units that would be equal to one-hundredth of a *koku*.

Takanori's pendulum clock was used in conjunction with other timepieces. Following his scheme of true local time, Takanori adjusted the pendulum clock every day at noon according to measurements made with the Sun Shadows Measuring Plate sundial installed in the Kanazawa castle. But the pendulum clock itself was used to adjust the "regular," non-astronomical mechanical clocks used by the bell keepers to announce the local time. The sound of the bell, in turn, enabled the citizens of Kanazawa to adjust their mechanical clocks accordingly.

Thus the order of the day at Kanazawa castle clock tower ran like this: At the sixth hour of the morning—as determined by the pendulum clock—the bell keeper would ring the bells. At the same time, relying on the measurements of the pendulum clock, he would adjust the weights on a foliot on the regular mechanical "tower clock" so that it measured day hours with a speed appropriate to the season. In the next hour the bell keeper would similarly adjust a smaller mechanical clock that could serve to correct the tower clock if necessary. Then, over the course of the next few hours, the bell keeper would listen for the signal of the two mechanical clocks and compare them to the pendulum clock and the astronomical sundial. When the two mechanical clocks sounded noon, the bell keeper would check the sundial, and when the sundial showed noon he would calibrate the pendulum clock and ring the bell.[23]

The different timepieces were measuring different kinds of time. All of them measured time, and all were supposed be as precise as possible, and we can even say that each one measured time for the purpose of sounding the bell. Yet looking at the particular function of each one of them, we realize that the kind of time each one measured, in fact, was slightly different. The sundial measured time that was based on the notion that there existed an astronomical correlation to geographic locality that revealed true local noon. The pendulum measured time that was "true" in its ability to capture miniscule divisions of temporal units. The "tower" mechanical clock measured human time that was true to the seasonal changes, and thus reflected the climate of the region. Takanori's assemblage of timepieces reflected the variety of timepieces used by various astronomers described in chapter 3—albeit all at the same time. Even though each one of these clocks served the same timekeeping system and the same temporal culture, their particular uses conveyed different perceptions of time measurement. Each one was entangled in a different set of associations.

In spite of the theoretical success of Takanori's timekeeping system, the reform was short lived. Takanori's notions of "true time" perhaps resonated with other scholars in the domain, but were not shared by the majority of people who lived in Kaga. They were not initiated into the astronomer's practices of observation, calculation, and representation; they did not necessarily associate their land with some kind of "true" time as Takanori did; they may have been vaguely aware of the famous Kaga-born scholar, Honda Toshiaki, but it is unlikely that they were familiar with the details of his thought. For them, the measurement of the "truthfulness" of time was its compatibility with the long-established patterns of work and leisure that the new system seemed to neglect. Kaga citizens found the absence of the distinct "addition" unit confusing, and complained about what they saw as the chaos that the new system brought to old and well-calculated work schedules. Less than two years after the implementation of the reform, its most powerful proponent, the lord of Kaga domain Maeda Norinaga, died, leaving Takanori defenseless against prevalent criticism. Following popular demand, the system of thirteen hours was reinstituted and remained in place almost until the Meiji reform. The modified temporal system still contained elements of the precision measurements instituted by Takanori, which were supposed to reflect the specific geographic location of Kaga. Yet the replacement system did not encompass Takanori's full vision of "true Kaga time"—only those elements that the majority of people could agree upon.

Cartography and Magnetic Declination

Revocation of the calendrical reform did not put an end to Takanori's involvement with time measurement. Nor did his conceptualization of time measurement remain static. Every time-measuring project he undertook contributed to the evolution of Takanori's associations with time measurement and time in general.

Takanori's perception of time measurement manifested in his dealing with the question of magnetic declination, which had arisen in the course of the surveying project of Kaga that took place between 1822 and 1828. The project was meant to emulate Inō's surveys of Japan, both in surveying method and in the quality of the maps produced. Magnetic declination was one of the problems Inō himself was struggling with. The fact that true north differed from magnetic north (the direction indicated by a compass needle) had been known in China from at least the eighth century, and had been discussed in Europe since at least the fifteenth century. By the mid-seventeenth century there had been an abundance of sources available in Japan—both from China and the West—that discussed magnetic declination and even provided concrete figures. Toward the end of the eighteenth century, however, Japanese astronomers realized that the numbers they found in foreign books not only were inconsistent but also did not match their own observations on the ground. At the beginning of the nineteenth century, Inō had become troubled by the lack of clarity about the actual value of magnetic declination, seeing this as an obvious impediment to the precision of his maps. He decided to settle the problem once and for all by conducting a series of observations in which he compared compass measurements of the north with his calculation of true north using astronomical observations. To his surprise, he was unable to detect any difference between them. He conducted additional experiments using different compasses, but magnetic declination was nowhere to be found. Puzzled by the results, Inō nevertheless did not doubt the validity of his calculations and measurements. Refusing to yield to any other astronomical authority, he simply concluded that both the Chinese and the Western writings must be wrong, their mistakes originating from the fact that neither could evidently produce a high-quality compass.[24]

Takanori struggled with Inō's assertion that magnetic declination did not exist. On the one hand, Inō's observations seemed to have proven his claim and prominent Japanese astronomers of the early nineteenth century, such as Takahashi Kageyasu, accepted the nonexistence of magnetic declination as a given. On the other hand, in the early 1820s, Kaga surveyors' observations suggested that magnetic declination may have existed after all. The information they had at hand was conflicting: in the 1770s Kaga surveyors established that the declination was about five degrees to the east, at the beginning of the nineteenth century Inō claimed it was zero, and now, in early 1820s, some observation suggested that there was a slight declination to the west.[25]

Takanori's 1824 observations, which were meant to finally (again) determine the value of magnetic declination, relied on the precision of an astronomical pendulum clock. First, Takanori needed to determine the precise moment of the local noon, the moment when the sun reached the southernmost point in the sky. In order to do so, Kaga surveyors observed the progression of the sun across the skies and, using the pendulum clock, recorded the two times when the sun was at exactly the same altitude—once on the east and then on the west of the meridian. Then, finding the middle point between these two measurements of the pendulum clock, they arrived at the precise number of oscillations that had been counted by the clock when the sun was at its southernmost point on that specific day. The group made this measurement for two days in a row and arrived at the difference between the number of oscillations that indicated the two consecutive local noons. Then, extrapolating from this difference, they determined the precise time of the local noon according to the clock on the following day. On the third day, they compared the predicted number of oscillations on the pendulum clock that should indicate astronomical local noon with the number of oscillations that the clock actually recorded at the moment when the sun crossed the meridian. According to their measurements, 77 oscillations separated these two points of time (the precalculated and the observed noon). Given that Takanori's astronomical pendulum clock oscillated 60,036 times per day, a value convertible to 360 degrees, the group calculated that this difference in time translated into a difference of 0.4617 degrees. So far, the only differences between Takanori's method and the ones used by astronomers of the mid-eighteenth century were Takanori's slightly more sophisticated means of determining local noon and his use of a pendulum clock. Takanori's clock was precise to almost a second, as opposed to the roughly 15 minutes that characterized the best clocks of the 1770s.

The following step, however, shows a crucial difference in his conceptualization of time measurement and the impact this conceptualization had on his results. For earlier astronomers who approached time measurement merely as a tool to quantify specific events such as noon, the difference above—0.4617 degrees—would serve as the actual angular difference between the true and magnetic north. Takanori, however, was trained by Nishimura Tachū in the tradition of Asada Gōryū. This meant that Takanori performed astronomical calculations by means of constructing diagrams and computing using spherical trigonometry. Thus, for Takanori, the value of 77 oscillations, or 0.4617 degrees, was not the sought value but just an angle in his diagram. This value allowed him to generate a further series of triangles needed to calculate the actual value of magnetic declination, which he eventually placed at 1.7513 degrees to the west.[26]

Determining magnetic declination, however, was skewed by Takanori's convictions concerning geographic reality. Thirty years later, he conducted the exact same measurement using the same equipment. To his surprise, his calculation of magnetic declination produced a value of almost three degrees to the west. He compared the compass measurements with the maps that his team had produced in the 1820s and arrived at the conclusion that the measurements matched the map. So what was the problem? Takanori's convictions about the inherent "reality" of geographical location as defined by its relation to celestial motions did not allow him to suspect that there may have been another reason for this conundrum. For him, the only possible explanation for the varying results was some kind of fault with the compass. He conjectured that perhaps it had been stored close to iron weapons, or that when it was re-magnetized with a lodestone something had gone wrong, or, perhaps that it was the nature of compasses that they changed their orientation over time.[27] For Takanori, the problem must have lain with the compass, because for him, the physical qualities of geographic location could not have changed.

In reality, however, all four measurements—the one in the 1770s, Inō's at the beginning of the nineteenth century, and Takanori's in 1824 and 1854—were correct (at least more or less, with regards to the first measurement). Magnetic declination was gradually changing, shifting slightly from east to the west over the course of one century.[28] Inō did not find any magnetic declination because in the early 1800s the magnetic declination in Japan was zero. And Takanori's own two measurements were correct too, reflecting the changing nature of the phenomenon. But Takanori's vision did not allow for that.

Insights from Europe

With the passage of time, Takanori was increasingly exposed to Western thought. Western science was far from new to him, as he had absorbed many of its basic assumptions during his early studies of the Asada school of astronomy. This early, mediated exposure to Western science provided him with building blocks for a new kind of knowledge that resulted from later encounters with foreign books and objects. In a sense, Takanori's engagement with the West was emblematic of early-to-mid-nineteenth-century Japanese attitudes toward Western practices and material culture. When exposed to Western books, images, and artifacts, he noticed those elements that corresponded to his own existing knowledge, concerns, and aspirations. We could say that Western culture simply gave him a new language and new tools to formulate thoughts and goals that he already possessed, and perhaps added several new associations with time measurement.

One of the major channels through which Takanori came to experience Western science and culture was a Kaga doctor, Kurokawa Masayasu, who was more than thirty years his junior.[29] Kurokawa was born in 1817 into a family of physicians practicing "Dutch style" medicine (*Ranpō*). It was called "Dutch" because it departed from the strictly "Chinese" style (*Kanpō*) of medical practice. However, even though practitioners of the "Western" style adopted an anatomical conception of the body and utilized specific Western medical practices, *Ranpō* was a hybrid that did not strictly replicate how medicine was practiced in Europe. More than anything, "practicing Western style medicine" in the beginning of the nineteenth century suggested an eagerness to learn more about how medicine was practiced by

Europeans. Thus, at the age of eleven, Kurokawa followed his father to Nagasaki in order to learn Western medicine under Philip Franz von Siebold at the latter's *Narutaki* medical academy. Unfortunately for Kurokawa (and even more so for everyone directly involved), the discovery of Kageyasu's maps and the subsequent banishment of Siebold from Japan interrupted these plans. Nevertheless, Kurokawa stayed in Nagasaki to study Western culture and medicine for more than ten years. During this time Takanori was already interested in Western books, and when Kurokawa returned to Kaga in 1840 to assume a senior medical position, Takanori enjoyed the benefits of Kurokawa's knowledge.

Once he was back in Kaga, Kurokawa proved to be an invaluable source for Takanori because he was familiar with Western approaches to scientific thought and practices and, significantly, because he could read Dutch. Thus, when Takanori noticed a particularly interesting sundial in a European book, he turned to Kurokawa for help.[30] This episode of acquiring Western knowledge through translation was a proactive step on Takanori's part. He did not simply receive the knowledge of a Western device from outside—it was not imposed on him in any sense. Rather, his existing knowledge of timekeeping and his scientific and managerial aspirations allowed him to notice an image in a European book and to identify it as promising for his own work. This promise was not inherent to the device itself, but rather emerged from Takanori's previous timekeeping activities.

The explanations that accompanied the image of the Western sundial, made accessible by Kurokawa, prompted Takanori to design another kind of a sundial. The official, learned, classical Chinese name he gave his new sundial was Device for Observation of Shadows (*Shikeigi*).[31] However, the device was quickly dubbed the more colloquial Find the Four (*Yotsu no shirabe*)[32] (fig. 6.2). The "four" were the four interconnected parameters of time, latitude, season, and direction that were so essential to every enterprise Takanori undertook. If one knew only two of the four parameters, observing the sun's shadow on the device would provide the remaining two. And knowing at least two was rarely a problem. Unless one had just awoken from a long sickness, he could be sure of the exact season (one of either Japan's twenty-four or even Takanori's forty-eight). For those staying in or around Kanazawa, the latitude would be known, and those who traveled would most likely have been equipped with a compass to determine cardinal directions.

Travel was indeed something that Takanori had in mind when designing his new sundial. The whole device was cut from a single piece of cardboard so that the separate parts could be assembled into a sundial or folded to fit a rather narrow wooden box, ideal for carrying around dur-

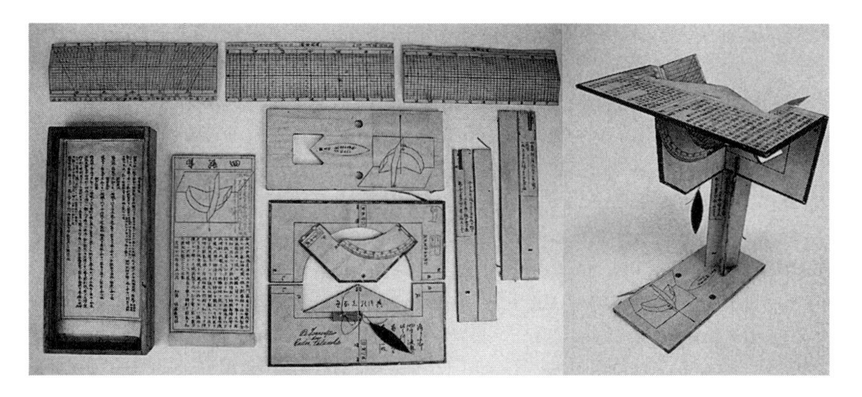

Figure 6.2. Device for Observation of Shadows or Find The Four (*Shikeigi* or *Yotsu no shirabe*). Ishikawa Prefecture Museum of History, Kanazawa. This device comprises parts made from paper and cardboard that can be disassembled and carried in a wooden box. The inscription in Dutch reads *"De Zonnewijzer door Endoo Takanolie"* [The sundial of Endō Takanori]. The three panels at the top of the image display both variable hours and equal hours written in Roman numerals.

ing travels. Thus the ideal user Takanori had in mind when devising the Device for Observation of Shadows was somebody like himself—a traveling scholar who needed to know the time for the purpose of recording his observations, and needed to know latitude and cardinal direction for the purpose of surveying.

In fact, the whole device reflected Takanori's perception of time and time measurement. It is true that the shape of the device was inspired by something he saw in a Western book, but at the same time it also manifested Takanori's associations with time measurement as well as his practical goals. Takanori took from the Western design those characteristics that confirmed his assumptions about time as being both localized by the sun's motion and geographically embedded in place through the measurement of latitude and cardinal directions. The sundial measured not just time, but Takanori's time.

Takanori's attitude toward time measurement manifested in his views on visual representation. By the 1820s Takanori had begun taking notes on various modes of depiction. These notes accumulated throughout the years and gradually provided him with enough material to publish a well-structured treatise entitled *The New Art of Depiction* (*Shahō shinjutsu*), completed in 1850.[33] Takanori was by no means an artist, as anybody reading the book could easily see. Yet his treatise, which explicitly states his views on aesthetics, is a treasure for historians of Japanese arts.[34] Here, however, we will focus on the intimate connection between Takanori's expertise in astronomy and surveying and his views on depiction.

In the book's introduction he described his motivations for writing it:

> Since antiquity the pictures drawn by sages were valued as objects of leisure, and those that imitate this style in the popular works of today are admired for their elegance. But for the matters used for governing, such as the shapes of the celestial phenomena, the body of the earth, the inside and outside of human body, the tools of literary and military arts, the materia medica of doctors (beasts, insects, fish, plants, mineral stones)—how come turning to the method of depicting the real, aspiring to capture the thing itself did not become popular?[35]

Central to Takanori's views was his perception of reality. The "real" that Takanori was concerned with was *shin,* which could be also translated as "true" or "correct."[36] In Takanori's opinion the true is the thing in itself, *arinomama,* free of decoration, interpretation, or personal judgment.[37] In other words, it is not human-dependent. This notion of reality is recognizable in Takanori's 1823 reform, in which he attempted to establish "true" Kaga time by liberating it from human convention and grounding it in astronomical and geographic certainty. At the same time, his attempt to arrive at the "true" time, as well as his experience trying to determine a value for magnetic declination, had taught him an important lesson—reality was extremely illusive, and even when one tried hard to describe it as it was, there always remained a gap or discrepancy. So how could one capture and reproduce this elusive reality while "avoiding the discrepancy [between the depiction and] the shape of the real thing"?[38]

The mode of depiction, according to Takanori, is defined by the approach and the method of the painter. One of the major distinctions he drew was between three methods, each one based on the assessment of different things: mind, things, and observation[39] (see fig. 6.3). In the first category

---

Figure 6.3. The three methods of visual representation according to Endō Takanori. *New Art of Depiction* (*Shahō shinjutsu*). 1842–1850. Japanese Academy of Science, Gakushiin Archive, Tokyo.
Taking the moon as an example, Takanori shows the difference between "assessment of the mind" (upper panel), "assessment of things" (middle), and "assessment of observation" (bottom). When depicting the moon using the "assessment of the mind" method, the artist appeals to the viewer's emotions and may depict an exaggerated landscape, in which, for example, the spots on the moon or the clouds around it may be drawn bigger or smaller than they actually are in relation to the size of the moon. When painting the moon according to the "assessment of things" method, the artist depicts the existing environment, striving to realistically show proportion. The "assessment of observation," however, presents the moon as measured and explored by astronomers using their instruments, devoid of framing by clouds or mountains.

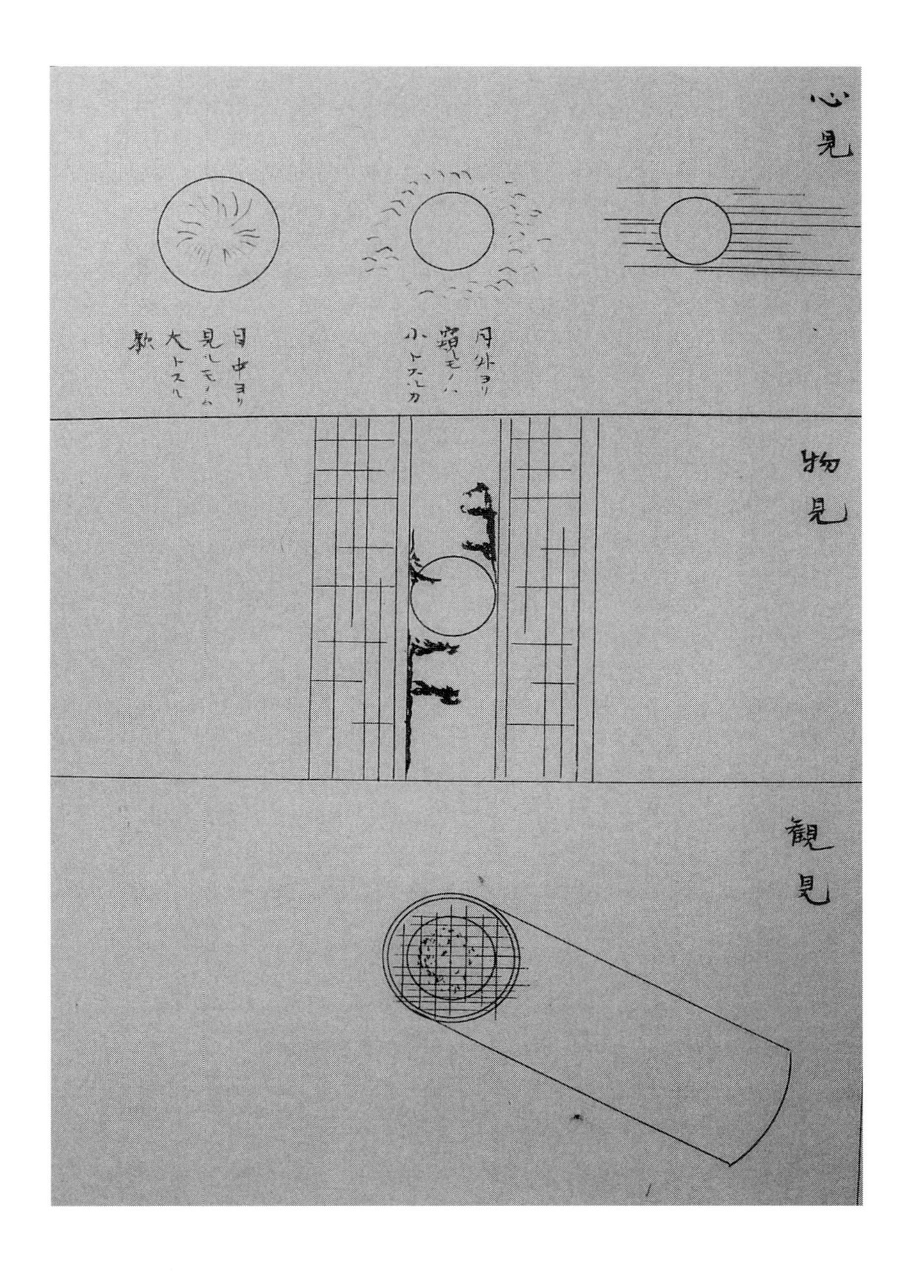

心見
目外ヨリ窺モノハ小トスルカ
目中ヨリ見ルモノハ大トスル歟
物見
観見

—a method of assessment of the mind—he put everything that he deemed elegant and pleasant to the eye, but useless "for the matters of governing." The second method—of the assessment of things—involved projecting three-dimensional reality onto a two-dimensional surface and disregarded "body" aspects such as thickness and weight, focusing only on the proportional representation of measured dimensions. This method served as the basis of the depiction of surveyed terrain, and thus was favored by Takanori for its suitability to governance.[40] The third method—of the assessment of observation—entailed a quantitative definition of reality that could not be directly measured, such as distant stars, for example. The eye can perhaps see them, but without the aid of a ruler, it cannot measure them, because the eye does not see in numbers. In these cases, according to Takanori, one must conduct measurements of arcs and degrees using instruments, such as sextants, theodolites, compasses, and microscopes, which allowed one to arrive "at the essence of the thing," expressed by numbers.[41]

Although Takanori incorporated in his treatise elements he saw in Western books, his theory was molded by his own practice. From today's perspective, it is easy to see in Takanori's treatise traces of Albertian rules of perspective and Baconian notions of instruments as an augmentation of human sense organs—notions that he had probably picked interpreting images in Western books. Yet he could only be "influenced" by Western books on drawing because he had already developed a perception of the world that was compatible with linear perspective. He learned how to see the world in this way through astronomical diagrams, by observing celestial bodies with a sextant, and by assessing landmarks with theodolites. He wrote:

> Astronomers begin their measurements by first determining the center of the earth. Surveyors determine the present position, or base. And whether you measure units of length, or numbers of things, or measure time, you start with the carpenter's square, the pointing of the finger, or the beginning of the day—they all start from one point. From there, you stretch a line. Astronomers stretch a line to the upper or the lower side of the sun, or measure to the east or west. Surveyors observe a landmark that falls in their sight, and stretch a black line to it, finding three lines with which one can measure angles. Whether it is astronomers determining diameters of the sun and the moon, or surveyors determining the length of an open field, it is a fine work making two lines and determining an angle between them. Having done that, astronomers explain the distance of the bodies from earth, and surveyors use trigonometry to determine the measurements of the surface.[42]

Without being directly and explicitly exposed to the theory of Western optics, Takanori had nevertheless absorbed some of the principles through a wide variety of Western observational, calculational, and measurement practices that had been integrated into Japanese astronomy and surveying. He did so by learning how to make calculations using diagrams, by observing stars with a sextant and landmarks with a theodolite and transcribing these observations as lines, by translating time into angles and triangles, by assessing the world with numbers, and more. Taken separately, any one of these practices could not have communicated the entire Western scientific worldview. Like the first European mechanical clocks, any one of these practices by itself would have been transformed into something that better fit local meanings and practices. But once integrated into Takanori's web of practical, material, and visual associations, Western practices became nodes that could be connected to other practices, providing additional conceptual ground that enabled access to additional foreign notions and over time generating a critical mass. All these practices together, enabled Takanori to extrapolate and often arrive at conclusions that were somewhat similar to those arrived at in the West. Thus, Takanori was enamored with linear perspective and hailed it as "true" because he extrapolated from his notions of time and space that were already linear, geometric, and numerical.

The Unlikely Tale of a Russian Clock

Takanori's life experiences also reflected the growing consciousness of seafaring, described in the previous chapter. Early in his life, Takanori encountered the world outside Japan through the (often somewhat fictional) writings of Honda Toshiaki. Yet over time he saw additional European books, used Western instruments, and heard stories about Westerners and the ways they conducted observations and measurements. As noted above, he found much that connected to his own work on Kaga time and geography, and he incorporated conceptual, material, practical, and visual elements of Western culture into his own web of associations. By the middle of the nineteenth century, he was familiar enough with Western culture that he could engage directly with European practices of science without the mediation of the Chinese Jesuit authors whose texts had introduced Takanori to Western science.

A culmination of Takanori's later engagement with Western timekeeping was his chance encounter with a Russian clock. In *Memoir of the Presented Clock and Its Use* (*Kenjō onjiki yurai narabini yōhō no oboe*), Takanori recorded the account of sailors who had presented him a Russian clock for his inspection. According to Takanori's record of their story, a merchant ship with a crew of ten had left Kaga in 1838 to trade with different domains of the Japanese archipelago. At the end of that year, as the ship sailed from Sendai on the Eastern side of the main island, the crew encountered the worst nightmare of Japanese mariners of the Edo period—a strong storm swept them into the open ocean. Takanori did not have to explain the full horror of this situation, as the predicted fate of castaways was well known at the time. Even if they survived the storm, the official government policy that forbade Japanese from going abroad would prevent them from ever coming back. Even worse, those who were swept to the north were doomed to spend the rest of their days among people whom the Japanese deemed "barbarians." In reality, by the 1840s there were a fair number of castaways who had managed to return to Japan after all, but the fear of living among barbarians and not being able to return home was nevertheless very present in the public mind.

Takanori's account describes the adventures of the Kaga castaways from their wreck in 1838 until their return to Kaga ten years later. The crew did survive the storm and soon after was rescued by an American ship, which dropped them off at the island Takanori called "Santoisu" or "Sandoisu."[43] About a year later, the Kaga sailors were picked up by an English ship that dropped them off at Kamchatka. They then traveled to Okhotsk on mainland Russia, and after a couple of months were put on a ship back to the other side of the ocean, to Svetskaya Sitka (current day Alaska, then the capital of "Russian America"). In 1843, the Russians took them back to Japan. By then, only six sailors remained. The captain of the Russian ship, or "*nachalnick*," as Takanori described him, repeating the Russian word uttered by the survivors, was "Adolf Karlovich," aka Arvid Adolf Karlovich Etholin (1799–1876), a Finnish Chief Manager of the Russian American Company. Etholin treated the Japanese sailors well, even inviting them for drinks in his inner cabin. Seeing what a great impression his clock made on the sailors, he ripped it off the wall, handed it to his visitors, and instructed them to present it to their lord when they got home. After several months at sea they reached the island that the Russians called Iturup (the Japanese called it Etorofu, and the local Ainu population called it Etuworop). The sailors took all their possessions, including the clock, and continued by foot to the nearest Japanese city. It took them a month to reach Matsumae, located on the southern tip of Ezo (Hokkaidō), but the ordeal was far from over. Since officially they were not supposed to return to Japan, they were sent to Edo for questioning. Only several years later were they granted permission to return to Kaga. Only four of the original ten made it back home by the end of 1848.

The clock—the reason behind Takanori's recording of the story—made it too. Surprisingly, the sailors kept the clock with them the whole time, dragging it back and forth across Japan over difficult terrain through all kind of weather. When they presented the clock to the lord of Kaga upon their return, it was broken. It was given to a local clock-maker for repair and when the work was completed the clock was handed over to Takanori, the local expert in time-related matters, to decipher the unfamiliar way the Russians told their time. Of the clock, Takanori wrote:

> Besides striking the twenty-four hours, this clock also struck every thirty minutes (*bu*). In the cities of the Russian state there are time-bells, which are placed inside buildings similar to [our] towers. The clocks would, of course, strike in the same way [as the bell]. Moreover, in addition to striking the twenty-four hours, one could also observe a movement of the two index-hands,[44] and by seeing their position immediately be able to know the hour. In terms of its use, there is a long and short index hand. Their length made appropriate, the long one makes 24 rounds, and the short one makes 2 rounds.[45]

Takanori's story of the Russian clock linked seafaring, international relations, and the Western method of keeping time. Unlike previous, mostly hypothetical theorizing about the role of timekeeping in international travel, this story had faces, and it had an object to support it. In order to bring his story to life, Takanori began his narration with a map (see fig. 6.4). On this map Japan was depicted as just a couple of pink spots near the bottom of the page, while a vast open ocean occupied the majority of the available space. The ocean was divided by the grid of longitude and latitude lines, but without a clock to measure their locations, one would be lost and wander from place to place like the poor castaways.

* * *

Takanori's engagement with timekeeping uniquely drew together his interest in astronomy, infrastructure, surveying, and modes of representation. It was also noticeably local and Kaga-specific in its character. Yet, at the same time, Takanori's associations with time, as formulated in his work, are also representative of wider transformations in the conceptualization

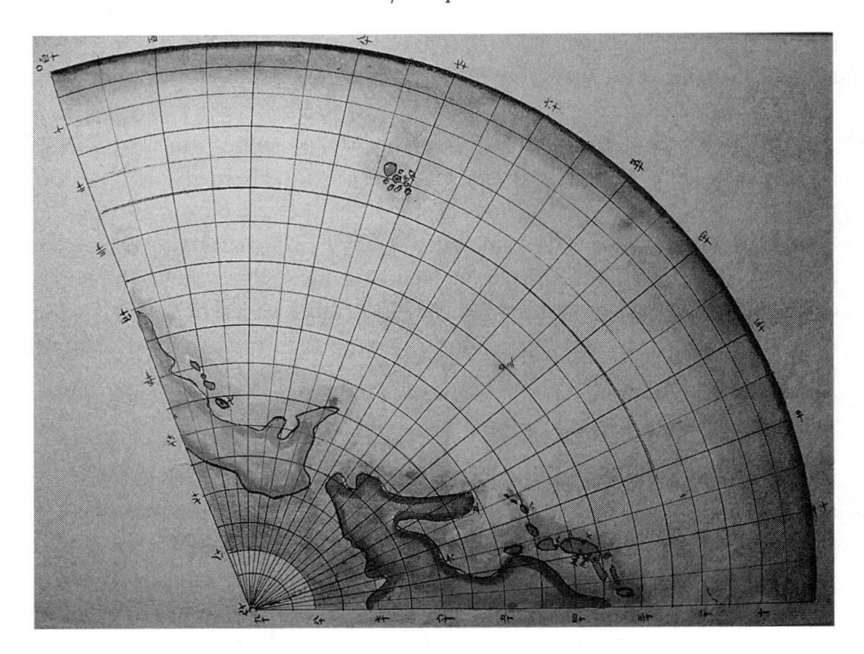

Figure 6.4. Endō Takanori, The opening page of *Memoir of the Presented Clock and Its Use* (*Kenjō onjiki yurai narabini yōhō no oboe*). 1849. Japanese Academy of Science, Gakushiin Archive, Tokyo. Takanori's narrative, which focuses on the efforts of Japanese castaways to bring a Russian clock to their lord back in Japan, is prefaced with a map of the vast ocean traversed by the sailors. In the story, Takanori's Russian clock becomes not just a gift from one high-ranking official to another, but a symbol of the interconnectedness of geography and a reminder of the importance of time measurement in navigation.

of time measurement that took place in Japan during the first half of the nineteenth century. Attuned to the evolving time-measurement practices of astronomers and geographers in Edo and in Kyoto, Takanori integrated them into his own endeavors, and thus incorporated their underlying associations into his own thought. Combining the various fields in which those time-related associations manifested themselves, he connected the dots and created a picture of time measurement that resonated with many scholars of his time.

Color Plate 1. A hanging clock (*kakedokei*) with two foliots. The Seiko Museum, Tokyo.

The digits on this dial are arranged counterclockwise (e.g., the digit indicating the hour of Sheep—the hour after the noon Horse hour—is on the left of the Horse digit). The dial itself rotates clockwise while the index hand remains stationary, pointing up. To know the time, the viewer needs only to look at the top of the dial, where the index hand indicates the current time. In the position shown, the index hand indicates that it is almost the hour of the Sheep.

Color Plate 2. A pillow clock with movable digits. The Seiko Museum, Tokyo.

Tokugawa Japanese used square wooden boxes with a padded surface to support their necks during sleep. After the Meiji period the meaning of the word they used for this item—*makura*—gradually transformed and came to indicate a pillow, namely a soft, cushion-like pouch. For early eighteenth-century Japanese, however, European box-like clocks looked exactly like the pillows they knew and used.

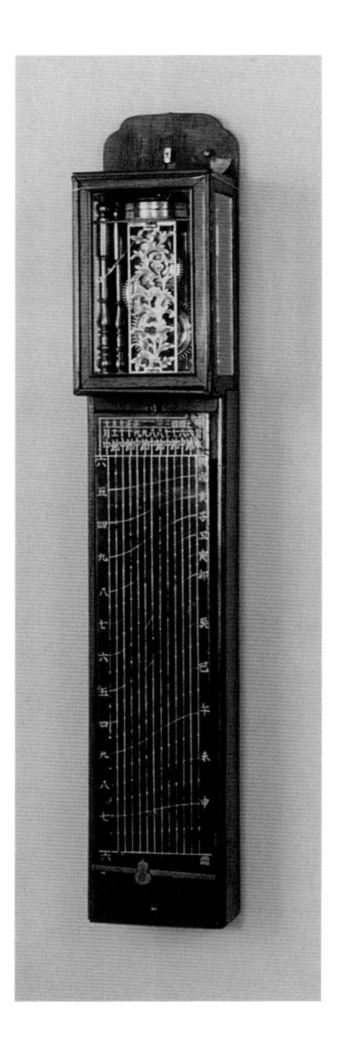

Color Plate 3. A "medicine box" clock (*inrō tokei*). The Seiko Museum, Tokyo.

This portable clock was worn tied to one's belt like a medicine box. The elaborate decoration of this example references a famous woodblock print by Suzuki Harunobu titled "Evening Bell of the Clock" (*Tokei no banshō*), from the series Eight Views of the Parlor (*Zashiki hakkei*). Its lavish case made this particular clock a luxury item, but many medicine box clocks were simpler in appearance. Photograph courtesy of The Seiko Museum.

Color Plate 4. A *shaku* clock with a graph-like dial. The Seiko Museum, Tokyo.

This clock mimics the graphs on paper sundials (fig. 2.8), as well as the pattern created when seasonal *shaku* plates are arrayed side by side (fig. 2.7). The x-axis of this dial corresponds to seasons, while the y-axis displays hours. In order to find the time, one slides the bob horizontally on the index rod (positioned here at the bottom of the dial) to the appropriate seasonal slot. The clockwork mechanism moves the rod downward from the sixth hour of the dawn at the top of the dial to the following sixth hour of the dawn at the bottom. Photograph courtesy of The Seiko Museum.

Color Plate 5. A round graph-like clock. The Seiko Museum, Tokyo.

Like the graph-like dials of *shaku* clocks, this dial combines seasonal variations in hour distribution into one graph, albeit a circular one. The clock mimics the illustration of seasonal changes in the length of days and night, in which the graph consists of concentric circles divided into one hundred units (see fig. 2.9). This clock is fully automated, and does not require manual seasonal adjustment. Instead, the index hand extends and retracts according to the season (shorter during the winter months, extending to the outer edge of the dial during the summer). To find the time, the user observes the location of the tip of the index hand, rather than the direction in which the tip points.

Color Plate 6. Inō Tadataka's pendulum clock. Inō Tadataka Memorial Museum, Katori. Photograph courtesy of Professor Nakamura Tsukō.

This clock was designed to count 59,504 oscillations of its pendulum per day under ideal conditions but variations of several oscillations per day were common. The upper dial displays single oscillations, the middle dial indicates tens of oscillations, and the lower dial counts hundreds. Two small windows at the bottom of the lower dial display a count of thousands and ten-thousands.

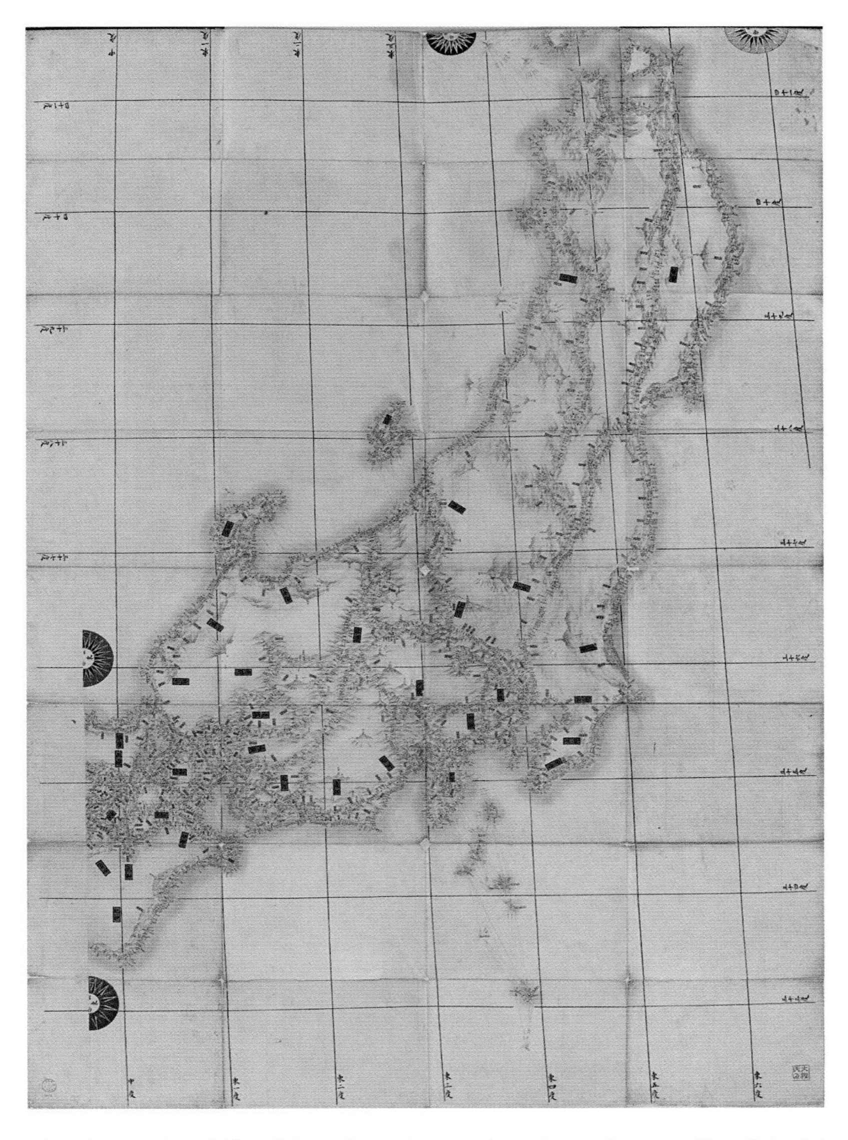

Color Plate 7. The middle of three sheets that constitute the smallest set of Inō Tadataka's Maps of Greater Japan. *Comprehensive map of the coasts and the lands of the Greater Japan. Small Map. Eastern Part of Honshū* (*Dai Nihon enkai yochi zenzu. Shōzu. Honshū tōbu*). Original size 245 × 164 cm. Tokyo Metropolitan Library.

Color Plate 8. Model of the Sumeru World (*Shumisengi*) built by Tanaka Hisashige. The Seiko Museum, Tokyo. Photo courtesy of Ōhashi Clock Shop.

This mechanical model of the Buddhist universe was meant to convince the observers that Buddhist cosmology is, in fact, plausible, as it could be recreated by mechanical means. The central structure in it is Mount Sumeru, each side of which is painted according to the four seasons (summer is gold, autumn is red, winter is blue, and spring is green). The dial at the peak displays the twenty-four mini solar seasons (*sekki* 節気). The disk hovering above the mountain is marked with the twelve "seasonal months" (*choku* 直), which were defined not by the waxing and waning of the moon, but by the annual rotation of the Big Dipper. Attached to this disc so as to orbit Mount Sumeru are spheres representing the sun (gold) and the moon (silver). The characters on the outer ring enclosing the white plane indicate the twenty-eight constellations ("lodges" *shuku* 宿) of Chinese astronomy. Displayed prominently on the model's front is a clock, which emphasizes the connection between time measurement and the structure of the universe and suggests that the universe itself is composed of clockwork.

Color Plate 9. Tanaka Hisashige's *Man-nen dokei.* Important Cultural Property of Japan. Deposited object from Toshiba Corp. Permanently exhibited in the Tokyo National Museum of Nature and Science.

This "forever clock" has six dials that display six different types of time—variable hours, European equal hours, the twenty-four seasons, astronomical units of *koku,* the date according to the sexagenarian cycle, and the date in terms of months and days. At the top of the clock is an astronomical model, simulating celestial phenomena observable in the skies above Japan. To create a mechanism capable of these myriad functions, Hisashige invented new shapes of gears, and manually shaped teeth in order to improve the mesh between gears. The clock also features panels ornamented with gold, enamel, mother-of-pearl, pressed rare woods, and brass engravings—all executed by highly skilled artisans.

Color Plate 10. Westernized Japanese clock. The Seiko Museum, Tokyo.

After the Meiji calendrical reform, this pillow clock with movable hour digits was converted to measure the twenty-four equal hours of the European system. Not only were the formerly movable digits placed equidistantly, but an additional dial with Arabic numerals was added on top of the existing one. Unlike the clock-makers who had previously modified Western watches for Japanese consumption, early Meiji-period masters did not alter the mechanism of the clocks they modified, but simply added dials with twenty-four positions. Despite its rotating dial, stationary index hand, and nonstandard Arabic numerals, this clock did actually keep time according to the Western temporal convention. However, by displaying the time in two types of units simultaneously, the clocks allowed early Meiji users to learn how to translate their experience of daily time into Western temporal units.

SEVEN

Clock-Makers at the Crossroads

At this point, an attentive reader might observe that all the transformations I've documented in the perception of timekeeping happened among a fairly small circle of people, one that only narrowly extended beyond an elite group of professional astronomers and geographers. But did any of these transformations resonate among the general public? Surely, the country was not headed for calendrical reform just because of several dozen highly specialized professionals. The answer to this question is a straightforward "yes and no."

Let us start with the obvious. The vast majority of people during the Tokugawa period understood neither the notion of mean time, nor the practical complexity of determining longitude at sea. Most were probably unaware of astronomers' practices, and even among more educated scholars interested in astronomy not many were particularly sensitive to the relationship between timekeeping and astronomical calculation. But the creation of a fertile environment for the reform of the temporal system did not require the whole population, or even the majority of it, to understand those practices. As we will see in chapter 8, it was that small circle of people who came to believe in the superiority of mean time who drove the calendrical reform of 1873. They hardly needed the approval of the general populace.

But this was not completely a top-down decision, as many urbanites in late Tokugawa Japan had gradually come to see the Western temporal system as advantageous. Their reasoning was not unrelated to that of astronomers, since a range of associations related to astronomical practices trickled into the popular imagination. This did not happen magically, by some kind of invisible force. Rather, there were specific people, go-betweens, who facilitated the dissemination of astronomical associations among the general public.

Those go-betweens were clock-makers, who produced both measurement instruments for astronomers and mechanical devices for popular consumption. In order to accommodate astronomers' needs, clock-makers had to learn about astronomers' goals and motivations. Nevertheless, astronomical instruments accounted for only a small portion of clock-makers' sales, which was comprised largely of devices for popular consumption. Having been initiated in astronomers' worldview, however, clock-makers disseminated astronomers' associations through their work—through the devices they made and through writings in which they discussed the technology they dealt with.

Karakuri

Central to the dissemination of associations and assumptions related to time measurement was an art that grew out of the clock-making industry—the mechanical art of automata.[1] The earliest known reference to automata is in a short verse from 1675 written by the famous Japanese novelist Ihara Saikaku:

Edo Harima, Takeda from Osaka,
amassing the wisdom of the Tang,[2]
crafted a spring-driven device, making it carry a tea cup.
The movement of the eyes and the mouth,
its manner of walking, extending its arms, bowing at the waist . . .
Just like a human being![3]

This short verse provides us a glimpse into the world of the mechanical art of *karakuri*, which was already thriving in the seventeenth century. Harima was the nickname of an automata master from Edo who was well-known at the time but who didn't leave as big a mark on the history of automata as the other person mentioned in the verse—Takeda Ōmi, a clock-maker who had a workshop in Osaka during the middle of the century. Owing to his mechanical skill, he was able to make extra money by constructing a variety of devices referred to in Japanese by the term *karakuri*. The etymology of the word is murky, and in most cases it was written using a phonetic alphabet. Broadly speaking, the word *karakuri* could refer to any kind of complicated device—a magnetic compass, a secret box that required one to solve a riddle, and, certainly, any kind of device that contained gears. *Karakuri* didn't necessarily have to be operated by gears, though—devices operated by an intricate set of strings were perfectly qualified, as well. In this sense, broadly applied, the word *karakuri* could be translated as "device" or "mechanism."

The automaton mentioned in Saikaku's verse—the Tea-Carrying Automaton—epitomized Tokugawa period automata.[4] The puppet, set in motion by a spring, was released by the host of a party and "walked" toward one of the guests carrying a tray with a teacup (or a sake cup).[5] Bowing his head, the puppet invited the guest to pick up the cup and drink. Once the guest finished the drink and returned the empty cup to the tray, the puppet turned around and walked back to the host.

Not unlike modern-day android robots, it was the high degree of likeness to human motion that made an automaton a sensational spectacle. Lifelike features served to testify to the mastery of the artist who made gears and strings emulate human movement. The idea that these automata manifested was that mechanical structures had the potential to re-create anything, even the complexity of human motions, and all that was required to turn this potential into actuality was the skill of the automata master.

In 1730, another clock and instrument maker, Tagaya Kanchūsen, published an *Encyclopedia of Automata* (*Karakuri kinmō kagami gusa*) that described dozens of strange types of automata. The pictures show *karakuri* in action exclusively, and Kanchūsen did not attempt to describe their inner mechanisms, reflecting, perhaps, his assumption about his readership's lack of interest in that aspect of the topic. The inner workings were intricate and complex, yet early eighteenth-century readers were seemingly satisfied with the spectacle produced by the mechanical art and evidently did not need to know *how* exactly the effects were produced (see fig. 7.1).

The shifting of popular interest in the inner workings of automata followed clock-makers' trade with Dutch precision instruments. At the time, a variety of instruments were being imported to Japan through Dejima. The instruments—sextants, telescopes, theodolites, and others—were not only carefully crafted but also purposefully designed to *produce* precision. Collectively, they prompted the emergence of the idiom "Dutch craftsmanship," *oranda saiku*, implying that it was especially precise. Japanese clock-makers created local versions of such devices and were proud of their ability to reproduce the "exquisite precision," *seimyō*, required for the task.[6] The instruments were, of course, created for astronomers' use, but nevertheless entered popular consciousness through popular science books written by amateur enthusiasts of "Dutch studies," or *Rangaku*.

One of the first Dutch devices popularized by *Rangaku* scholars was the

Figure 7.1. Encyclopedia of Automata (*Karakuri Kinmō Kagamigusa*). 1720. Waseda University Kotenseki Sogo Database of Chinese and Japanese Classics. The left panel shows an automated fan, the right a writing automaton. Although not all the devices depicted in this book were likely built, there is abundant evidence that the two depicted here actually existed.

electrostatic cabinet. After several months at sea, the first one to make it to Japan arrived broken, but it was repaired by the famous polymath Hiraga Gennai.[7] The electricity-producing device provoked quite a stir. For some scholars it appeared as a possible alternative means to administer moxibustion, while others were drawn by the spectacle of sparks flying out of the hair of a person who held onto the handle. It was promptly described in a book by Gotō Godōan titled *Dutch Tales* (*Oranda banashi*), published in 1765. The *erekiteru*, as the device was dubbed in Japanese, was featured in one of only three illustrations in the book, the other two being of a Dutch man and a Dutch woman. Interestingly, the image and the discussion of the device immediately preceded the pages that discussed a variety of clocks brought by the Dutch or made in Japan. The *erekiteru*, thus, was perceived both as a pinnacle of mechanical wonder and as belonging to the same category of ingenious machines made with exquisite precision as *karakuri* puppets and clocks.

Some twenty years later the device again appeared in a popular book, similarly titled *Various Dutch Stories* (*Kōmō zatsuwa*).[8] The author was Morishima Chūryō, a literary student of Hiraga Gennai and younger brother of the shogunal physician Katsuragawa Hoshū, who was one of the doctors involved in the translation of Dutch anatomical tables that was published as *The New Book of Anatomy* (*Kaitai Shinsho*) in 1774.[9] Owing to his peculiar position at the intersection of the literary and professional worlds, Chūryō both was acquainted with the latest scientific novelties arriving in Japan aboard Dutch ships and had the literary talent to craft a national best seller. Describing the functions of the different parts and how they connected, Chūryō fueled the already growing popular interest in mechanics. Amateur readers still admired the effects created by mechanical devices, but now they wanted to know more—how the inner apparatus was able to produce effects by means of connected gears. Examining the inner construction of mechanical devices, readers rapidly became aware of a fact long known to all *karakuri* artists—that in the heart of many automata beat a clock mechanism.

Reverberations of the Kansei Calendrical Reform

The realization that automata were driven by clocks resulted in the association of all mechanical devices with clocks,[10] as exemplified by the *Illustrated Manual of Curious Machines* (*Kikōzui*, or *Karakurizui*, in alternative reading), first published in 1796.[11] Its author, Hosokawa Hanzō, had been born into a high-ranking house in Tosa, in Shikoku. As expected of someone of his status, he was classically educated. However, in addition to the regular classical curriculum, he also studied mathematics and astronomy. In 1791 he traveled to Edo to work with government astronomer Yamaji Saisuke, a colleague of Takahashi Yoshitoki.[12] Since his teacher belonged to the group of astronomers later responsible for the Kansei calendrical reform, Hosokawa too became involved in the project, though more on the technological than the mathematical side.

Having worked on the construction of astronomical observation and measurement instruments, Hosokawa decided to write a book about clockwork principles, including the dynamics of falling weights, the elasticity of springs, and magnetic forces. The book was titled *The Record of Machines Reflecting the Heavens* (*Shatengiki*), suggesting that the material phenomena he encountered in clock-making were manifestations of universal forces.[13] The book, perhaps too professionally oriented, failed to reach a wide audience, prompting Hosokawa to write the *Illustrated Manual of Curious Machines*, which immediately became a best seller. The *Manual* was written in accessible language and included numerous pictures, yet it was essentially a similar attempt to propagate the idea that mechanical clockwork echoes universal forces. Hosokawa died shortly after finishing the book, and the preface was completed by none other than Morishima Chūryō. It stated the following:

> When you continuously witness what it takes to construct curious machines, it leaves an impression on your heart. Touching a thing, you learn how to utilize mechanisms—noticing the way a fish moves its tail in the water [you] make a wheel. [When you] look [at the way it] moves with its fins left and right, [you] make things like oars. Long ago, Chuko Kongming designed a "wooden oxen and running horses" while watching his wife make dolls.[14] Takeda Ōmi discovered the essence of *karakuri* while watching a child play with the sand. Although this book is indeed like child's play, I hope it would awaken the minds of those who look at it with consideration.

The book positioned the clock as a tool for understanding the world. According to Hosokawa, since technology imitated the principles found in the surrounding world, by learning about mechanical technology, one also learned about how things worked in reality. And since the most basic form of mechanical technology was the clock, then in order to understand any technological (or natural) principles, one had to first learn how clocks worked. Consequently, Hosokawa opened the *Illustrated Manual of Curious Machines* with a chapter dedicated exclusively to a variety of clocks, only afterward moving on to discuss the mechanisms of automata (see figs. 7.2 and 7.3).

The *Illustrated Manual of Curious Machines* was not written for people who wanted to *make* automata but rather for those who wanted to know *how they worked*. The *Manual* probably provided inspiration for aspiring automata makers, some of whom indeed eventually constructed mechanical devices. But to be clear, no one could possibly have learned how to build a clock just by reading this book alone. The *Manual* neither described the latest technological developments, nor provided the technical details essential to the actual construction of the mechanical devices it described.[15] But it did go into great detail explaining the names, shapes, and functions of each and every gear and bolt. These would have been trivial to any instrument maker, but for educated amateurs it was an insight into the workings of the marvelous machines, and thus, as the introduction implied, into the workings of the universe.[16] Reading through the book, the reader witnessed the virtual process of construction without actually acquiring the practical knowledge necessary to build an automaton. Hosokawa's book revealed how a device made of brass could perform functions like measuring time, bringing tea, descending stairs, or writing calligraphy. Hosokawa knew

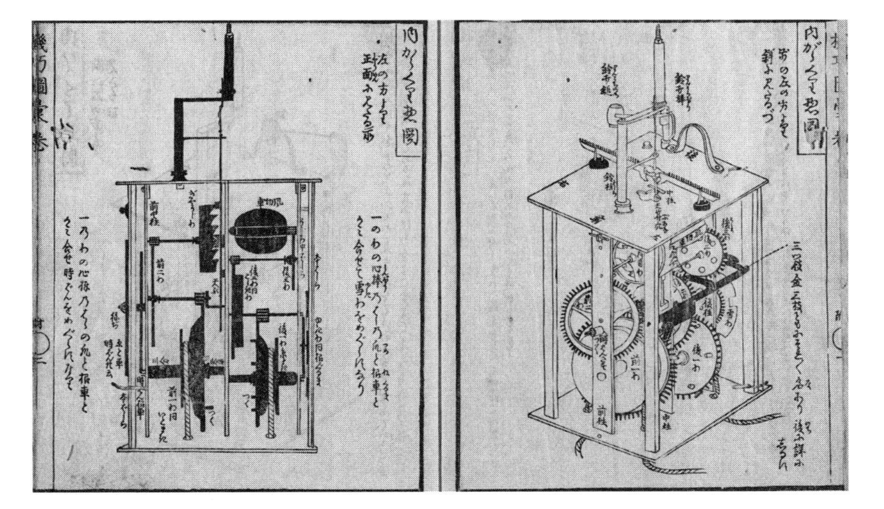

Figure 7.2. *Illustrated Manual of Curious Machines* (*Kikōzui*, or *Karakurizui*). National Diet Library Digital Collections. The author claimed that in order to understand *karakuri* one must study the mechanics of clockwork. For this reason, he dedicated the first volume of his manual to clocks, discussing aspects of design and construction but offering few details—such as specific dimensions—necessary to build a clock.

what his audience wanted—in a section that explored the *shaku* clocks, he made a point of indicating that such clocks had a glass window on their upper side "through which one looks at the movement of the balance."[17] In the sections on automata, he revealed which parts made the legs walk and the arms bend. It was his explanation of how gears could move in such a way that they produced seemingly human functions that made *Illustrated Manual of Curious Machines* into a best seller. The clocks and the automata described in the *Manual* were demystified; it was the exquisite skill of the master that remained a mystery.[18]

The images in the *Manual* certainly contributed to its appeal. Curious readers were tempted to copy the images, following the construction of the device through drawing.[19] The anonymous author of the *Illustrations of Timepieces*, for example, borrowed some of Hosokawa's images, but only rarely mentioned any technical information.[20] Contrasting the seeming indifference to technical details with the painstakingly detailed (yet somewhat amateurish) illustrations, one gets the impression that it was the joy of investigating the structure, of identifying the parts of the clock, that motivated the author to produce this work. The depiction of the disassembly of a clock and related mechanical investigation, followed by the copying

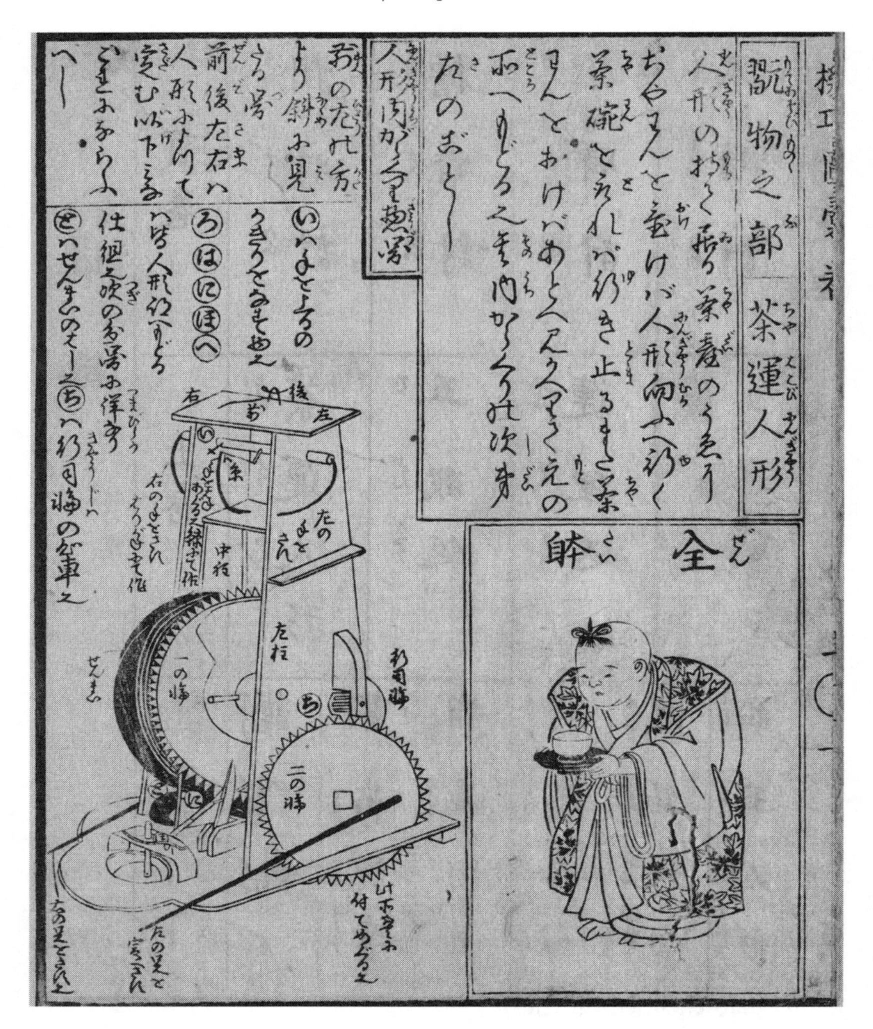

Figure 7.3. Tea-Carrying Automaton in the *Illustrated Manual of Curious Machines*. National Diet Library Digital Collections. The second and the third volumes of the manual are dedicated to humanoid and animal-like automata. Tea-carrying automata, mentioned by Ihara Saikaku as early as 1675, were—and still are—considered paradigmatic *karakuri*.

of the pictures from *Illustrated Manual of Curious Machines*, amounted to a virtual reconstruction of a clock in its own right. Such amateurs were driven neither by profit nor by a childlike fascination with the mysterious "inside." Rather, they were attracted to the exquisite product of human skill and ingenuity—mechanical motion.

Clock-makers like Hosokawa Hanzō were living links between astronomers and the general populace, and served as a channel for the spread of tidbits of knowledge about the latest astronomical developments. Clock-makers were the ones astronomers turned to with requests for new instruments—and not just timepieces, but also any observation or measurement instrument that required precise execution and a practical knowledge of working with brass. There were instruments designed by astronomers themselves—such as the *shaku* clock of Asada Gōryū or the pendulum clock of Hazama Shigetomi—which needed to be built according to specifications that reflected the designer's particular intent. There were also instruments, such as sextants and theodolites, which were imported from Europe through Dejima and needed to be adapted to a format familiar to Japanese astronomers. Whatever the case, clock-makers had to learn and understand astronomers' motivations and assumptions about astronomical observations and calculations, as well as the assumptions implicitly embedded in the instruments themselves, local and foreign alike. But since clock-makers did not receive a government salary for their efforts, they were eager to apply their newly acquired knowledge to make further profits in the general market.

On the most immediate level, clock-makers disseminated astronomical knowledge by selling observation and measurement instruments to lay amateurs. Having already designed all the different parts and invested in building customized equipment for the casting of brass components, they were happy to manufacture more of the same instruments for sale. There was a steadily growing market for these instruments. Various domains carried out survey projects independent of the ones sponsored by the central government. But as letters to Takahashi Yoshitoki show, there were also rich amateurs who dedicated their free time to the seemingly sophisticated and useful practice of surveying and making astronomical observations.[21] We can also easily recognize Inō Tadataka's instruments in the advertisements published by clock-makers' shops (see fig. 7.4).

Clock-makers' apprentices popularized what they had learned from the astronomers in their own work. Take, for example, Kume Eizaemon Tsūken.[22] Having become a clock-maker's apprentice at age seven, at eighteen he was sent to study under Hazama Shigetomi, who described him as "absolutely brilliant in building machines. He takes a look at the pendulum clock, the sextant, and the compass used in my family, and is able to make them from the scratch."[23] Back home in the Takamatsu domain in 1806, Tsūken was entrusted with a surveying project, for which he made a set of instruments that replicated Shigetomi's equipment. In doing so he

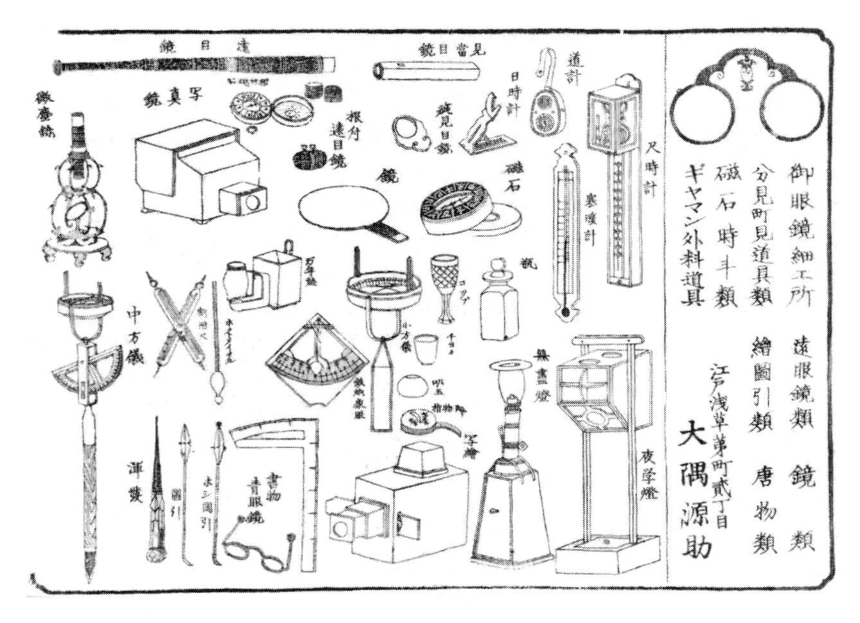

Figure 7.4. Advertisement for measurement instruments.
Late Tokugawa period. The Seiko Museum, Tokyo.
Alongside imported objects such as glass cups are a *shaku* clock, compasses, rulers, and a variety of theodolites, similar to those used by Inō Tadataka and his successors. Although originally designed for use by specialists, instrument makers soon marketed instruments to a wider public.

facilitated the spread of the latest astronomical practices, bringing to Takamatsu what had been previously restricted to central astronomical circles. In 1808 he was ordered to join forces with Inō Tadataka, whose survey expedition passed through Shikoku island, and to integrate his own measurements with those of Ino's nationwide survey.[24] Later, while continuing to work on a variety of mechanical devices, Tsūken used his mechanical expertise to create technologies for a globalized world. Unlike the optimistic Honda Toshiaki, however, Tsūken had a much grimmer outlook, focusing not only on shipbuilding but also weapons.

Astronomical activity surrounding the Kansei reform also contributed to the rise of scientific translations of Western literature. Astronomers needed translations of Dutch astronomical literature and worked closely with Dutch interpreters, educating the latter in the principles of astronomy. In one instance a young Dutch interpreter, Baba Sajurō, was brought to Asakusa observatory to familiarize himself with the body of knowledge he would need to translate. Sajuro went on to translate not only books on astronomy but also ones on medicine, glass-making, meteorology and thermometers, and, of course, European clocks.[25]

Such collaboration between astronomers and interpreters resulted in a wave of books describing not only astronomical instrumentation but also the cosmology that served as the basis of European astronomical practices. For professional astronomers, heliocentric theory was old news, and the transition from the Tychonian model of the universe to the Copernican one was perceived as a natural progression—a far less dramatic change in astronomical practice than the transition to the elliptical orbits of Kepler's laws. For amateurs, however, even for those who studied astronomy, the Copernican model was deeply thought-provoking.

The link between Western cosmology, Western styles of representation, and Western astronomical instrumentation was most clearly articulated in the work of the famous Dutch studies scholar Shiba Kōkan.[26] Kōkan is most widely known for his exploration of European techniques of visual representation. The first to produce copper etching in Japan, he was also famous for his landscapes, which utilized linear perspective. Around the turn of the eighteenth century, he began to focus on a very particular aspect of perspective and representation, that of the skies and the cosmos. The geometry on which the heliocentric view of the universe was based was the same as the one that had guided artists in their development of the rules of linear perspective. The same laws of optics that underlay visual effects in painting also served to explain such phenomena as eclipses and parallaxes. And instruments such as sextants that allowed astronomers to make observations were, of course, instruments of vision. Arriving at these conceptual connections, Kōkan was inspired to publish a series of richly illustrated books on Western—and specifically Copernican—astronomy, in which he also described in detail the variety of instruments employed by astronomers to investigate the heliocentric universe.[27]

Clockwork Universes

But how influential were these books? Did they manage to convey their message? Did they convince their audience to accept Western astronomical theory and practice in its popular form? Finding in the sources an explicit "yes" once, twice, or even several times would still not add up to an afirmation that could be considered representative. Instead, we find evidence of the prevalence of Western cosmology and its mechanical association in the words of somebody who actively wanted to say "no."

Fumon Entsū was a Buddhist monk who made it a personal crusade to fight Western cosmology.[28] As a student at the Myōyō temple of the Nichiren sect around the end of the eighteenth century, he was already troubled by the rising popularity of Western astronomical theories, which contradicted Buddhist cosmology. Entsū himself was not swayed by Western theories, sticking to the view that only the spiritual wisdom of Buddhist doctrine could offer insight into the real structure of the universe. But he could not fail to notice that books propagating Western theory were quite convincing. "Convincing" for Entsū did not necessarily mean "true," however. Therefore, instead of fighting Western cosmology on ideological grounds, or trying to convince others to accept the superiority of Buddhist doctrine, Entsū decided to adopt a different strategy, embracing the very methods and characteristics that made Western theory "convincing."

In Entsū's view, what made books about Western astronomy convincing and appealing was the ability of Western sciences to geometrically outline, graphically represent, and realistically depict cosmological ideas in images, maps, celestial and terrestrial globes, and armillary spheres. Western cosmology was convincing because it could be re-created in a good "model," which, in Entsū's definition, was a "form" that "represents" and "confirms [something]."[29] The solution to this problem seemed obvious to him—he needed to create a realistic model of the Buddhist world that would correspond to both unarguable mathematical data and observed experience, and which would convince people of the validity of Buddhist cosmology.[30] In 1810, in a treatise titled *The Book of Astronomical Phenomena in Buddhist Countries* (*Bukkoku rekishō hen*), he described a model of a universe based on traditional Buddhist cosmology.[31] At its center stood a trapezoidal Mount Sumeru, surrounded by mountains, seas, and lands.[32] The sun and the moon, together with the other planets, all circled the mountain, creating what appeared to humans as years, seasons, months, and days. For Entsū, it was this cosmological model of Mount Sumeru, rather than the one used by astronomers of his time, that should have provided the basis for the calculation of the calendar.

But why discuss Entsū and his fantastic models in a book on the gradual Westernization of Japanese time? His mission was to undermine the professional astronomy of his day and to force the use of the "Indian"—i.e., Buddhist—calendar in Japan.[33] Not only did he object to the Western heliocentric model, he even rejected the geocentric model used in classical Chinese astronomical treatises such as the influential thirteen-century *Season Granting Calendar*. Entsū was a revisionist, and, indeed, in many twentieth-century history books he was represented as reactionary.[34] He was not taken particularly seriously by his contemporaries. The Buddhist establishment refused to accept his theories, and he had to frequently change his sectarian affiliation. Among astronomers, he was a joke. Although Inō Tadataka wrote a treatise refuting his model, many other astronomers thought that engaging in any kind of discourse with Entsū was simply beneath them.[35]

The interesting thing about Entsū, however, was that although he explicitly rejected Western scientific theory and cosmology, he actually accepted numerous assumptions and associations that were tacitly embedded in Western scientific practices. His explicit objection to cosmological models prevalent among astronomers of his time only highlights the fact that he too was immersed in the web of associations with time measurement shared by many of his contemporaries. Explicitly, he rejected the relation between timekeeping and the Western cosmological model, but implicitly, he repeated many of the assumptions underlying the Western practices adopted by early nineteenth-century Japanese astronomers.

Let us examine some of those assumptions that shine through his revisionist façade. First of all, his justification for Buddhist cosmology was based on the contemporaneous astronomical methods of mathematical calculations, observational data, and experiments. For example, he provided different interpretations of experiments proving the sphericity of the earth and criticized existing calculations of the planet's mass.[36] Though he denied the validity of a particular Western model, Entsū nevertheless wholeheartedly accepted the assumption, held by Japanese astronomers, that validity is established by mathematical calculation.

His description of the Buddhist universe revealed another underlying assumption, according to which realistic visual representation testifies to the validity of the model. Initially, "realistic" for him meant "adhering to mathematical parameters" and representable in a diagram. However, the word Entsū used to describe his "model" (*gi* 儀) was also used to denote "instruments" or "devices," such as for example in the name of the pendulum clock—*suiyōkyūgi*. The choice of the word indicates that perhaps from the very beginning his intention was to create a working model of the Buddhist universe.[37]

The opportunity presented itself in the late 1840s, when Entsū witnessed the technological wonders of the Edo period's leading *karakuri* master, Tanaka Giemon Hisashige. Hisashige was born in 1799 in Kurume, the son of a tortoise-shell artisan.[38] He was skilled in *karakuri* making from relatively early in his life, and most of the surviving Edo automata are his creations. It is worth dwelling on a few of these designs, as they reflect the particular early nineteenth-century appreciation for mechanical art. Hisashige left some parts of the mechanism exposed so that the viewer could see the movement of the gears that set the humanlike automaton in motion. Moreover, sometimes he added parts unnecessary to the function of the mechanism that instead served to emphasize the mechanical nature of his art. Take, for example, his arrow-shooting puppet. The puppet, sitting on a pedestal, reaches with her hand to grab an arrow from a stationary quiver.[39] She then notches the arrow on the bowstring and draws the bow. Moving her head, the puppet then appears to be aiming, after which she releases the arrow, which flies and hits the bull's-eye of an appropriately placed target. Apparently satisfied, the puppet then reaches for another arrow and repeats the process. The tiny movements of the hands and the head make the puppet display not only a remarkable physical capability (skillfully shooting the arrow) but also mental states (intensity, concentration, satisfaction). Yet Hisashige was not content with this display alone. At the bottom of the pedestal on which the puppet sits, he placed another figurine, a much smaller puppet of another child who appears to be cranking a wheel that sets the larger doll in motion. In actuality, the small puppet's movements are created by the same set of gears as those of the arrow-shooting girl on the pedestal, a fact that viewers are easily able to discern. The tongue-in-cheek gesture suggests that what appears to be alive was in fact set in motion by another human, which is clearly itself an automaton, all of which, of course, is actually set in motion by a set of gears constructed by one very real human—Tanaka Giemon Hisashige, or Karakuri Giemon, as people came to call him.

In a sense, automata epitomized late Tokugawa utilitarian philosophy (*jitsuyōgaku*)—the gears *visibly produced an effect,* they made the puppet *do something,* and this *something* was usually a useful activity: writing, shooting a bow and arrow, bringing tea. Notwithstanding the fact that their only functionality was their capacity to entertain, they *looked* useful. But like many of his contemporaries, including Kume Tsūken, Hisashige did seek to make his art useful by also creating mechanical devices with practical capacities. At the age of thirty-five, he moved to Osaka and earned money by demonstrating, and sometimes even selling, devices that he had invented, such as an automatic oil-feeding "inexhaustible" lamp or a foldable lantern for travel. He continued working on mechanical inventions throughout his life and during the last years of the Edo period he designed rifles, steam engines, a telegraph, keyless locks, a universal screw cutter, a lathe for cutting elliptical sections, an automatic tobacco cutter, a candle-making machine, an oil press, bicycles, water pumps, and much more. Early in the Meiji period, he moved to Tokyo to establish an electric workshop, which would evolve into Tokyo Shibaura Electric Company—Toshiba.[40]

Having seen Hisashige's wondrous automata and his numerous "useful" inventions, Entsū approached him and asked him to produce an equally convincing and useful-looking model of the Buddhist universe.[41] True to Entsū's Buddhist cosmology, the Model of the Sumeru World (*Shumisengi*) Hisashige produced was a clockwork-based mechanical device in which the sun and moon circled above a trapezoid mountain painted in four colors to represent the different seasons.[42] Reminiscent of an orrery, it provided the viewer with the positions of the sun and moon as well as information about the tides and lunar phases. Traveling through various domains, Entsū lectured about his observationally supported and mathematically defined Sumeru World, and accompanied his lectures with demonstrations of one of the Models he ended up ordering[43] (see color plate 8).

Entsū, intentionally, was propagating his view of the universe as surrounding the hourglass-shaped Mount Sumeru. Meanwhile, his model conveyed the set of notions, already prevalent in popular Japanese science literature, that something that worked mechanically must be true, that at the core of the universe there was a clockwork mechanism, and that the motion of the mechanical clock was directly linked to the movement of celestial bodies.

Entsū's success inspired one of his followers, Sada Kaiseki, a monk from the Jōdoshinshu Honganji school of Buddhism, to utilize additional mechanical models to support his preaching about Buddhist doctrine.[44] With the help of none other than Hisashige, Kaiseki created a Model of Both Observed and Real Phenomena (*Shijitsutōshōgi*) as an answer to the large-scale Model of the Sumeru World (*Shumisengi*), which didn't represent the earth as it was commonly shown on maps of the world[45] (see fig. 7.5). Kaiseki's model showed the earth immersed in the vast sea surrounding Mount Sumeru and its position relative to the sun. In Kaiseki's Model of Both Observed and Real Phenomena, "observed" referred to cosmography supported by contemporaneous astronomical theory, while "real" denoted the cosmography according to the Buddhist worldview.[46] Consequently, the device did not aim to overturn the accepted heliocentric cosmography. Rather, it attempted to show that the celestial and terrestrial phenomena described and measured by astronomers were true, yet provided only a partial picture of the much larger Buddhist universe. The Model of Both Observed and Real Phenomena positioned the audience at an objective

distance from which it was possible to observe the heavenly rotations relevant to the human world but also the general motion of the sun and the moon above Mount Sumeru.[47] Similar to Entsū's model of the universe, Kaiseki's model had a visible dial of a mechanical clock, but going one step further, the dial was positioned not on a side but directly at the center of the device. There were obvious mechanical reasons for building the model using a clockwork mechanism, but there was no such reason to include a clock face, let alone to position it at the heart of the model. Its inclusion, for both Entsū and Kaiseki, clearly offered a message—mechanical clocks are representative of the orderly and constant workings of the universe. The validity of each model of the Buddhist universe was implied by its mechanical structure, suggesting that mechanical possibility is indicative of the "real" structure. Yet the fact that both models incorporated a real working clock tells us something more—the model had to abide by the physical rules guiding and enabling the motion of a mechanical clock. The clock, thus, became a symbol of the regularities in the natural world, a working microcosm of the universe.

Hisashige himself did not buy into Buddhist models of the universe. All the while he was working on the mechanical representations of Buddhist cosmology, he was studying Western astronomy, geography, and navigation. Already in the late 1840s he moved to Kyoto to study under a scholar of Dutch studies, Hirose Genkyō, who also happened to be his brother-in-law and a good friend.[48] There, he not only deepened his knowledge of astronomical theory—the kind that was supported by the leading Japanese astronomers of his time—but also learned about astronomical practices, and specifically, about the importance of time measurement to astronomy.

Perhaps as a balance to a model he made possible but did not believe in, Hisashige decided to create one that reflected the knowledge he held

Figure 7.5. Model of Both Observed and Real Phenomena (*Shijitsutōshōgi*). Tokyo National Museum of Nature and Science. Photograph courtesy of Professor Suzuki Kazuyoshi. This mechanical model of the Buddhist universe was created to convince the viewer that the universe according to Western cosmology is, in fact, a tiny part of the larger Buddhist universe. In this model, Western cosmology is equated to "observed" phenomena and represented by the Earth's four seasonal inclinations, seen as the four globes on the central surface. The Buddhist, or "real," universe is represented by the planetary model at the top of the device, as well as by the surface on which the globes "float," which represents the celestial ocean that surrounds the mythical Mt. Sumeru. The model positions the viewer at a distant, Archimedian point, from which one can view how the "observed" universe fits inside the "real" one. Testifying to the model's validity, a variable hours clock dial wraps around the central rod supporting the "real" phenomena above, its index hand attached to the "observable" structure mounted above the globes.

dear. This would be the ultimate timepiece, the one to incorporate all the known principles related to timekeeping and answer every time measurement need. The device in question was a state of the art timepiece that combined mechanical ingenuity, elegant design, and rare, expensive materials. But more importantly, Hisashige embedded in it the characteristics he associated with time measurement, crystallizing a particular moment in the history of Japanese timekeeping. Hisashige's clock was a material manifestation of the process in which a network of material, visual, and practical associations with timekeeping qualitatively shifted. His device shows how certain associations born out of idiosyncratic astronomical practices managed to work their way into the popular imagery.

The device was called *Man-nen dokei,* which literally translates as "ten thousand year clock," but connotes the meaning of "forever clock."[49] The name referenced the fact that it could go without winding for an extended period of time. An engraving on one of the internal plates noted that it should be wound once a year (somewhat more frequently than the ten thousand years promised by the name).[50] The clock represented the optimistic—and somewhat naïve—belief that there could and should be a universal device that would keep time correctly for eternity. And confirming this belief was Hisashige's certainty about the astronomical grounding of the temporal system.

The "forever clock" acknowledged that there were numerous ways to measure time (see fig. 7.6). Designing what he imagined as the ultimate timepiece, Tanaka included six separate dials, one for each different kind of time. The first of the six dials showed the hours of the day according to the "lay" Japanese system of variable hours. This dial had divided hour-digits (*warigoma*), similar to those of the pillow clocks described in chapter 2. However, unlike any other pillow clock, the digits were designed to change their position automatically, reflecting the length of hours during the various seasons. The second dial—which, by contrast, had to be manually adjusted—showed the twenty-four mini-seasons of the solar year. The third dial had two index hands—one rotated daily and indicated the variable hours arranged along a hundred-*koku* scale; the other had a weekly rotation and indicated the days of the week. The fourth showed the date according to the sexagenarian cycle, while the fifth showed the month and the day of the month.[51] Finally, the sixth dial was a European clock with a short index hand that rotated twice a day, measuring twenty-four equal hours; a long index hand that measured sixty minutes each hour; and an additional small dial that rotated through sixty seconds every minute (see color plate 9).[52]

Figure 7.6. Flier from 1851, advertising Tanaka Hisashige's "forever clock" (written as 萬歳自鳴鐘, which could be read either as *bansai jimeishō* or *mannen tokei*). Tokyo National Museum of Nature and Science. The ad is boasting the six dials, showing six kinds of time, as well as the astronomical model above. Hisashige had to advertise his clock to attract viewers who would pay for looking at this wonder, and hopefully a potential buyer.

The design conveyed the sense that all the various modes of time measurement were just different manifestations of the same core time embedded in the central mechanism. The six dials were arranged as six faces on a hexagonal column, on top of which stood a mechanical model, albeit not of the whole earth, but only of the Japanese archipelago. The model showed a map of Japan on a grid of longitude and latitude lines and encircled by two overhead loops representing the meridian and the celestial equator. There were two small spheres, one red and the other silver, that rotated around the whole structure to imitate the apparent motion of the sun and the moon in the skies above Japan. All those different functions were set in motion by the same mechanism.

The astronomical reality of the "forever clock" was the one defined by early nineteenth-century astronomers (see fig. 7.7). The red ball representing the sun reached an altitude of 55° at the time of the equinoxes, 78.5° on summer solstices, and 32.5° on winter solstices—values that correspond to the geographical location of Kyoto. The model also represented the sixth hour of dawn (*akemustu*) and the sixth hour of dusk (*kuremutsu*) according to the Kansei calendrical system devised by Takahashi Yoshitoki during the last decade of the eighteenth century, which placed the sun's position at those hours at 7.3611 degrees below the horizon—a value based on the calculation of the position of the sun at equinoctial time at the latitude of Kyoto. There were no numbers on the longitude and latitude grid, but one of the longitude lines run through Kyoto, similar to Inō's maps. In addition to motion that represented the daily rising-and-setting, east-to-west path traversed by the "sun," it also moved one degree to the east each day, representing the average orbital rotation of the earth. The "moon," too, moved 13 degrees to the east along its own path to represent its changing position relatively to the fixed stars. The red sphere representing the sun rotated with a period of 23 hours and 56 seconds, corresponding to the earth's axial rotation period with respect to the stars.[53]

In 2004, a team of experts took the device apart in order to examine its mechanism, revealing the extent to which Hisashige invested in minute mechanical adaptations in order to represent the astronomical specifications described above. He devised special gears with different numbers of external and internal teeth to represent the apparent motion of the sun and the moon, invented other "insect-shaped" gears to allow for the automatic adjustment of variable hours, hand shaved every tooth on every gear, calculated the lengths of controlling cranks, and much more.[54]

Hisashige's "forever clock" conveyed early nineteenth-century Japanese astronomers' assumptions about the relationship between the physical

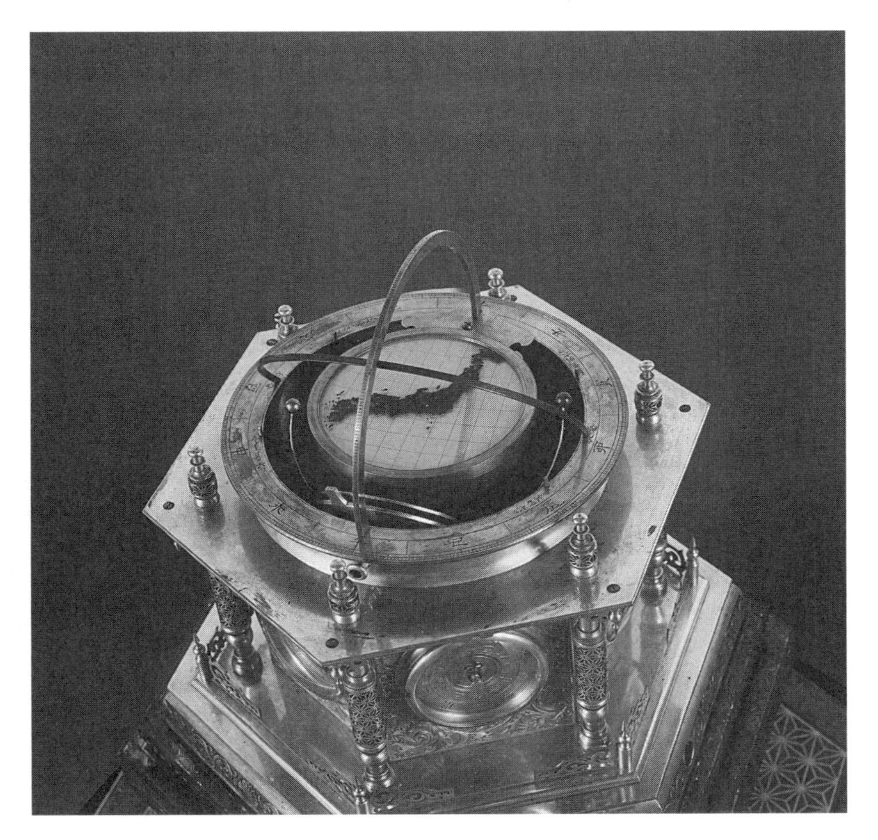

Figure 7.7. Tanaka Hisashige's Myriad Year Clock. Tokyo National Museum of Nature and Science. Photograph courtesy of Professor Suzuki Kazuyoshi. The top of Hisashige's clock features a multi-axis mechanical model of the motion of the sun and moon in relation to the Japanese archipelago (shown on the round plate at the center). Hisashige placed the map of Japan on the grid of longitude and latitude lines as calculated by Intttō Tadataka. The motion of the sun and moon (represented by the small spheres on either side of the map) is calibrated to Japanese astronomers' calculations. This mechanical model was connected to the central mechanism that controlled the clock dials below, projecting the idea that apparent celestial phenomena and the various kinds of time experienced by people were in fact all governed by the same clockwork mechanism.

reality of the universe, time measurement, and mechanical technology. It validated astronomical principles by showing that they physically and visually worked, as set in motion by a clockwork mechanism. It provided a three-dimensional visualization of the interconnectedness of the Japanese islands through their temporal and geographic links to the origin of Japanese time—Kyoto. Moreover, it also conveyed that time had many faces and could be measured in a variety of ways, but that all of them were just particular manifestations of the same mechanism that also determined the motion of celestial bodies. And, following this line of thought, the device suggested that mechanical clocks had an ontological significance as manifestations of the inner working of the universe—the Western dial showing the average motion of the earth, the Japanese dial reflecting the astronomical underpinnings of seasonal changes.

The "forever clock" played a significant role in further disseminating and reinforcing those assumptions and associations. While constructing this ultimate clock, Hisashige aspired to a perfection and beauty that were not only technical but also aesthetic and material. The dials contain jewels, the plates are made of rare lacquered wood with pearl inlays, enameled natural scenes were painted by famous masters and executed using rare materials, and many ornamented parts are made of precious metals. The clock, thus, did not just represent an astronomical model, it also cost an astronomical sum. Although Hisashige made a special effort to advertise the clock through copperplate prints, its enormous cost deterred even the richest and most technology-crazy potential buyers. To make up for the cost of its production, he thus resorted to traveling with it around the country, charging money for the viewing of this technological wonder. Unlike Inō Tadataka's maps or the pendulum clocks and chronometers that were accessible to only a narrow strata of scholars, the "forever clock" was displayed in public, spreading both its explicit and implicit messages among the general populace.

Reading Western Clocks

Clock-makers also played a crucial role in spreading the appreciation of *Western* clocks, and helped the general public make sense of the Western timekeeping system. As mentioned in chapter 2, the Japanese clock-making industry expanded over the course of the eighteenth and nineteenth centuries, both producing new designs for Japanese clocks and modifying imported ones, molding them into the forms familiar to the Japanese. At the beginning of the nineteenth century—about the same time as Japanese astronomers began eyeing European chronometers as their preferred time-measuring devices—Japanese consumers gradually began to develop a taste for unmodified Western watches.

One characteristic of Western watches that appealed to nineteenth-century Japanese consumers was the presence of Arabic numbers. With the growing acknowledgement of the importance of commerce and the subsequent rise in the status of mathematics, numbers came to be seen as rational and objective, and were taken to be a direct expression of economic aptitude. Scholarly discussion of economic as well as of moral affairs came to include more and more numbers that served as testimony to a given description's objectivity. Foreign numbers were especially regarded as a symbol of efficiency due to the widespread utopian vision of the West as efficient, well-organized, productive, and lacking wastefulness.[55]

The mere inclusion of Arabic numbers on the face elevated the value of a clock. Even clocks designed in Japan contained Arabic numerals. The numbers 1, 2, 3, or some other combinations of numbers, began to appear on a small dial that became a common feature of *shaku* clocks (see fig. 7.8). The dial had absolutely no function besides displaying the rapid movement of one quick hand, assuring the owner that the clock indeed worked. The tiny hand measured neither minutes nor seconds, and the Arabic numerals had no significance besides making a fashion statement. In the same manner that European consumers demanded "oriental" characteristics in their furniture, Japanese consumers wanted to see "European" exotics.

This obsession with Western numbers tapped into the confusion about the term *jishingi*, which was used to refer to both marine chronometers and pocket watches, as described earlier. Consequently, the possession of an unmodified European *jishingi* was sufficient to cast the glow of sophistication on the owner, even if he or she did not understand how to read time off the foreign dial.

This was not to the liking of professional astronomers, who saw this fashion as cheapening their professional pursuits. Using a pseudonym, Shibukawa Kagesuke wrote a short essay, "Pointing out the ignorance about pocket watches" (*Shōchū jishingi jimō*). In it, he complained that most people took these recently imported miniature watches and "just played with them for fun, looking at the structure of this new technology, as if it was not a device meant to show the daily time of heaven's movement."[56] His correction of "misconceptions" about pocket watches was perhaps too difficult to digest, since his own understanding of time measurement was grounded in observational and calculational astronomical practices, which were not at all familiar to the general populace.

The ones who did manage to explain Western watches to the public were the clock-makers, who had to "change [the European watches] into Japanese style, and make a new display, in order to allow people to know the hours and their exact divisions."[57] To do so, they needed first to learn the Western timekeeping system as well as the correlation between this system and the inner mechanical structure of the watch. Having learned how

Figure 7.8. A *shaku* clock with Arabic numerals. The Seiko Museum, Tokyo. This device offers an example of the fashionable use of Arabic numerals during the late Tokugawa period. What appears to be a dial here is not a timekeeping apparatus but rather an ornamental element located above the vertical *shaku* dial (not shown). Although the index hand does rotate, the dial itself is divided into thirty segments, marked only with three equally distributed Arabic numerals, 2, 3, and 9, which do not appear to serve any function other than the aesthetic one.

to adapt the Western mechanism to suit Japanese needs, clock-makers were in a much better position than the astronomers to explain Western timekeeping conventions to the general public.

In a book titled *Investigation of the Clock Dial* (*Jimeisho jiban kō*), Fujimura Heizō explained the rationale behind the apparently nonsensical Western clocks. The book was published by the same famous clock-making workshop of Tōda Tōzaburō that produced measurement instruments for Edo-based astronomers. This suggests that Fujimura himself was affiliated with the workshop, and was in an excellent position to translate the Western system into Japanese terms. Instead of tackling the core theories underlying the Western system, Fujimura addressed the issues that made this system incompatible with his readers' understanding of timekeeping. The Western hours, he warned, "were not the same hours as Japanese ones, and only during the equinoxes were they comparable."[58] There were twice as many of them in the West, making them much shorter than the Japanese hours. What was worse, except during the equinoxes, it was not clear how much shorter they were exactly. Japanese hours were determined by the dawn and dusk, so that the Rabbit hour *always* started with the dawn, and the Rooster hour *always* started with the dusk. But Western hours behaved erratically, totally independent of the dawn and dusk. Fujimura saw his task as conveying to readers that the Western clocks were not "broken," but simply operated according to a system in which dawn and dusk did not have so much importance (see fig. 7.9).

Fujimura was just one among several authors who attempted to explain the Western manner of reading clocks long before the calendrical reform of 1873. Others created "translation tools" that were meant to assist Japanese owners of European watches in reading the time off their devices. Treatises like *On the Movement of the Three Hands* (*Sanshin Hatsuei*)[59] and *An Outline of Western Watches* (*Jishingi Teiyō*)[60] provided detailed tables, comparative pictures, or even paper models.[61]

Ogawa Tomotada's *On Time Measurement with Western Watches* (*Seiyō jishingi teikoku kassoku*), for instance, included conversion tables that he claimed were made through astronomical observations.[62] The tables were divided into seasons on one axis and hours on another. In each cross-section there were two values: first, the numbers indicated by the hands of a Western watch (Ogawa did not even call those "hours," but rather "rotations"); second, the time indicated in conventional Japanese units. Consequently, his readers could take the tables, identify the current season, then look at the position of the index hand of the Western watch, and know what time it was in local terms. For example: "in mid-September, at the

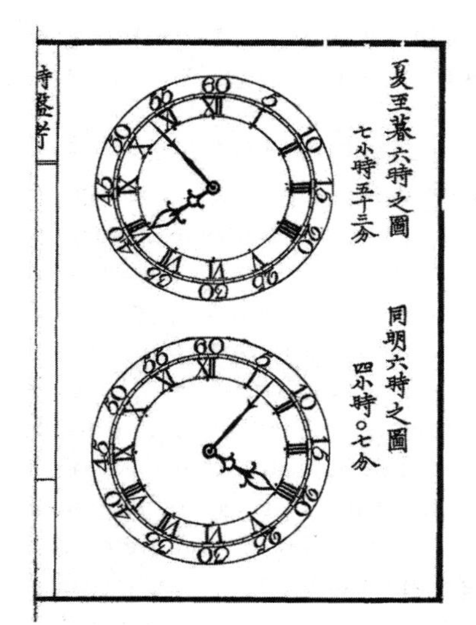

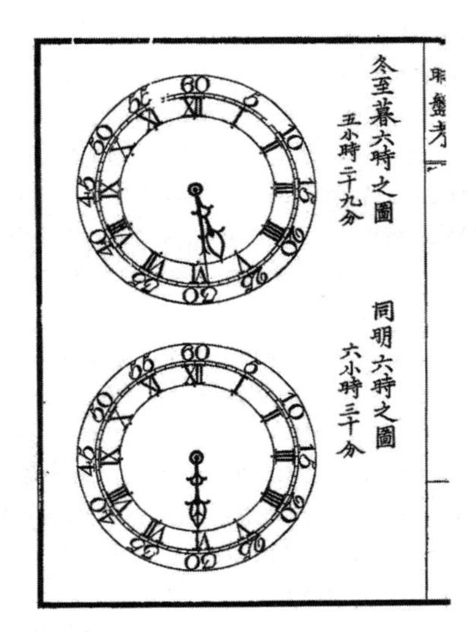

Figure 7.9. Image showing the positions of index hands on a Western clock dial at the sixth hour, which marks the beginning and the end of the day on the winter and summer solstices. Fujimura Heizō, *Investigation of the Clock Dial* (*Jimeisho jiban kō*). National Diet Library, Tokyo. The author points out the apparent absurdity of the Western system, in which the times of dawn and dusk change with the seasons. The dial to the upper left depicts the position of index hands on a Western clock at the sixth hour of dusk during the summer solstice, while the image below shows the position of hands at the sixth hour of dawn during the same season. Images on the right depict the corresponding positions of index hands during the winter solstice. Note the impossible position of index hands on the bottom right, which likely derived either from the author's perception of the bottom of the dial as the starting point of the clock, or from a misconception that the position of hands at the bottom of the dial must mirror that at the top.

eighth hour of the day, looking at the entry that matches the eighth hour, I have written down on the right in red, 'two rotations,' on the left side . . . , nine *bu*. Therefore, when it is two rotations, and nine *bu*, you know it is the eighth hour. And at the sunset hour, in the entry of 'six,' I have written in black on the right side, 'six rotations,' on the left, 'seventeen.' Therefore when it is six rotations and seventeen *bu*, you know it is the sixth hour of the sunset. All the others are the same."[63]

Moreover, Ogawa offered a method to estimate the correlation between the two systems even without looking at the tables. First, he advised to remember the correspondence "as if there was no seasonal differences." Then, one should remember by how many minutes Western units grow or shrink relative to the Japanese hours, and add or subtract minutes accordingly. Thus, for example, "during the first month,[64] a daytime half-Japanese hour is one rotation and two minutes, and nighttime half-hour is one rotation minus two minutes; therefore one should add two minutes to every day hour and subtract two minutes from [every] night hour. Therefore, a nine-and-a-half hour of day would be 'equal half an hour,' plus one rotation (of the short hand) multiply by two (minutes of the long hand), and so you get 1 and 2 minutes; from nine-and-a-half to eight: 'equal half an hour,' then two rotations multiply by two minutes and you get 2 and 4 minutes."[65] Ogawa dismissed concerns that this system was overly complicated. All one needed to do, he explained, was remember how to determine the changes from season to season, which was done by translating the difference in the degree of the earth's movement relative to the sun into time units of hours and minutes.[66]

Ogawa's tables created the sense that both timekeeping systems were grounded in the same astronomical reality. His astronomical basis for this was actually quite dubious. In his tables, hours always "grow" or "shrink" by exactly two minutes each hour, and although he mentions the fact that seasons are not symmetrical, the seasonal change is always described in three-minute units. Moreover, during the equinoxes, the sixth hour of the dawn and the sixth hour of the sunset fell exactly on round Western hours, perfectly reflecting the popular theory while ignoring the messy astronomical measurements. The numbers were rounded into neat units so that they could be conveniently used by the popular reader, who did not care about minuscule differences in measurement of time. Ogawa made the numbers usable and workable. Yet we should not dismiss his astronomical basis as a mere pretense. By pointing to his expertise in astronomy, Ogawa provided human time with universal validity. Without resorting to lengthy explanations concerning the difference between local solar time and mean time, Ogawa grounded human timekeeping conventions in celestial motion. His readers hardly got the whole picture of the astronomical reasoning behind the Western watch, but they understood that there was a connection between understanding Western time and understanding astronomy.

* * *

Clock-makers proved to be vital to the spread of new conceptualizations of time measurement. In order to satisfy the demands of their astronomer customers they needed to learn about astronomy; later, they disseminated their newly acquired knowledge both through popular writings about mechanical technology and by mechanical devices they themselves produced that embedded astronomical imagery. Working outside of highly professionalized astronomical circles, they were able to communicate novel associations with clockwork technology in a way that resonated with popular culture. Through the work of clock-makers, the general populace first came to appreciate the intricate structure of moving automata, and later recognized the central role of clockwork mechanisms in their movement. Soon, clocks came to be seen as a model for the functioning of *any* intricate structure, such as the universe itself. Even stringent opponents of mathematical astronomy, such as Entsū, who utterly rejected not only the heliocentric model but also the geocentric model of classic Chinese astronomy, wholeheartedly accepted the assumption that the universe is a built on a clockwork structure.

And having developed a strong association between clocks, in general, and a kind of ontological truth-value, early nineteenth-century Japanese also gradually came to value Western watches in particular. They did not need to understand the practical aspects of time measurement at sea, nor did they need to comprehend the astronomy and mathematical principles that dictated the differences between solar and mean time. But they did recognize that Western pocket watches, *jishingi*, were discussed in significant relation to international commerce, and thus came to associate those timepieces with progressive and utilitarian values. And here, clock-makers stepped in once again. Tackling all the apparent oddities and ambiguities of the Western approach to telling time, clock-makers gradually paved the way to comprehension, and even appreciation, of Western timekeeping conventions. By the mid-nineteenth century, without understanding the reasoning of astronomers, yet having gotten a glance at their conclusions through the work of clock-makers, the Japanese urban public had thus come to appreciate clocks in general, and Western clocks in particular, and was curious to learn and even try to use what was previously seen as an odd and nonsensical European time-measuring system.

EIGHT

Western Time and the Rhetoric of Enlightenment

So far, we have seen how both scholars and lay people of Tokugawa Japan gradually began associating Western-style timekeeping with advancement and sophistication. For astronomers and navigators the emergence of such associations was a matter of the evolution of calculational practices. Lay people, who got a glimpse of astronomers' concerns through the work of clock-makers, came to appreciate Western clocks as intricate mechanical devices associated with celestial movement. But fascination with Western clocks is still not enough to explain why in 1873 the Japanese government decided to convert the whole country to Western-style timekeeping. True, astronomers and navigators benefited from measuring time using European chronometers, but why would their preferences dictate the way the country as a whole measured time? After all, Tokugawa period astronomers had always used their own idiosyncratic systems of measuring time but so far the general population had taken no notice. The people on the streets of Edo and Osaka, on the other hand, may have become fascinated with European watches and timekeeping, but they still had their own system that worked just fine. This system was malleable and could have incorporated those aspects of Western watches that Japanese clocks did not possess, such as minute and second hands. Why, in the middle of the nineteenth century, was tinkering with designs suddenly seen as inconvenient and undesirable? What motivated the wholesale replacement of the existing timekeeping system? So far, we have seen how, in the first half of the nineteenth century, people gradually became predisposed to view the Western timekeeping system favorably. In this chapter I will focus on explicit, rather than implicit, expressions of this sentiment. Outlining the discourse on calendrical reform, I will describe how Western temporal conventions came to be seen as useful and superior while the Japanese system of measuring hours came to be perceived as burdensome. My research on the period leading up to the 1873 calendrical reform reveals that the decision to do away with variable hours was not rooted in any specific, practical problem. However, existing associations with both the variable hours system and Western watches came to acquire different meanings when situated in the political climate of the early Meiji period. The associations of Western clocks with astronomy and navigation came to bear increased importance in the context of the Meiji ideology of Westernization and enlightenment. The associations with variable hours and the clocks that measured them, on the other hand, were now linked to qualities deemed to be "backward" by Meiji intellectuals.

Creating Inconvenience

Scholars had begun raising the possibility of changing the calendrical system already by the end of the eighteenth century. Unlike astronomers, these scholars were not concerned with the correctness of the astronomical algorithm that served the existing calendrical system. Rather, they directed their ire at the luni-solar structure itself, and mostly at the final product of calendrical calculations—the printed calendar that was distributed annually by the government. At the end of the eighteenth century, musings about calendrical reform were distinctly utopian in nature, reflecting a Zeitgeist that called for widespread social reforms.

Already during that period scholars had begun expressing disdain with what they perceived as vulgar superstitions in the calendar. In 1789, Matsudaira Sadanobu, the de facto ruler of Japan at the time, wrote to a prominent Confucian scholar, Nakai Chikuzan, asking for advice about calendars.[1] Sadanobu had recently assumed his position as senior government councilor and shogunal regent and was planning his famous Kansei reform, which included the reform of the calendar carried out by Takahashi Yoshitoki (discussed in chapter 3). Sadanobu solicited Chikuzan's opinions concerning the state of the calendar and asked what potential changes would benefit society. Chikuzan was not able to help with the actual astronomical calculations of the calendrical algorithm—those were left to the professional astronomers. However, he did have some ideas about the final product—the annually distributed printed calendar. In particular, Chikuzan commented extensively on "the middle section" of the printed calendar, which included divinational information indicating auspicious and inauspicious cardinal directions as well as the dominance of particular deities.[2] The information in the "middle section" of the calendar was supposed to suggest which activities—building a house, signing a contract, holding a celebration, etc.—were favorable on a specific date. [3] In his reply to Sadanobu, Chikuzan did not spare harsh words about this part of the printed calendar, describing it as comprised of "ignorant fables, which greatly damage the public, and easily mislead the ignorant people of this realm, making it hard to enlighten them."[4]

Several years later, Honda Toshiaki would sound similar criticism and call for the replacement of the existing calendar with what he perceived to be the Western one. It is perhaps unsurprising that this admirer of all things Western was the first to consider adopting the European calendarical system. In his *Tales of the West*, Toshiaki praised Western knowledge of the three fields most important to him—astronomy, geography, and navigation—and suggested adopting the calendarical system that was based on knowledge accumulated across these fields.[5] Like his general description of Europe (discussed in chapter 5), Toshiaki's description of the Western system strikes the modern reader as naïve at best and ill informed at worst. Not only did he claim that the lengths of months did not vary in the West, but he also did not distinguish between the Julian and the Gregorian calendars and attributed the beginning of the Western year count to the reign of a Roman Emperor (without specifying which).[6] Toshiaki's description of the "Western" calendrical system reveals the items that were most troubling to him about the existing Japanese system—the disparity between the solar seasons (the *sekki*) and the lunar months, the fact that the arrangement of the calendar varied from year to year, and most of all, the inclusion of auspicious and inauspicious indicators, which he described using the colorful noun *idiocy*.[7]

The quest for a perfect calendrical structure that would reform the ills of society was also a trope in the writings of Nakai Riken, Chikuzan's younger—and equally famous—brother.[8] In addition to his extraordinary classical education as a Confucian scholar, Riken also studied astronomy with Asada Gōryū, and even produced a kinetic paper model of a heliocentric universe to be used for educational purposes. In 1801, just a few years after the publication of Toshiaki's *The Tales of the West*, Riken published his vision of a perfect calendrical system in a treatise whose title might be roughly translated as *A Calendar Invented by the Yellow Emperor in His Dream Country* (*Kashoreki*).[9] The Yellow Emperor, or Huangdi, was a mythical Chinese sage who, according to legend, ruled during the third millennium BCE and was credited with establishing the principles of cosmology, medicine, astronomy, and agriculture. By invoking this mythical figure, Riken indicated that he was striving to create a perfect calendrical system. Like his elder brother, he denounced the divinational aspects of the existing printed calendar, which he similarly referred to as "ignorant fables." He also viewed attempts to combine calculations of lunar and solar motion as cumbersome and inelegant, and thus based his calculations on solar motion alone. He was displeased that the number of days in the year did not reflect the number of degrees in a circle, which by that time—and under Western influence—was conventionally set at 360, not the 365 dictated by some traditions of classical Chinese astronomy. He thus proposed a year that was 360 days long, divided into four equal seasons. With its balanced divisions, Riken's system appears more than anything else like a straightforward demand for regularity and mathematical harmony.

Qualms about the clumsy complexities of the luni-solar calendrical system were echoed by another famous scholar and a student of Riken, Yamagata Bantō.[10] Like Riken and Toshiaki, Yamagata looked to the West for his model of a progressive society, and saw Western scientific practices as essential to achieving his social goals. In his *Instead of Dreams* (*Yume no shiro*), he decried the fact that Japanese calendars were overly complicated and praised the Western temporal system. In his words, the beauty of the Western calendar lay in the fact that "the seasons and the big and small [months] are known even to girls and infants."[11]

All of these writings share similar concerns: that the existing luni-solar system was too convoluted and inelegant in its irregularity and was therefore too complex for commoners to understand; and that existing annual calendars contained instructions for divinational practices, which scholars deemed detrimental to popular morality. Although the quality of calendars was measured by the quality of the astronomical algorithm on which it was based, in the discussions leading to the Meiji reform, criticism was not leveled at inadequate astronomical calculations but rather at the relationship the general population was perceived to have with certain aspects of the existing calendar, particularly the provisions for divinational practices. Reformers not only wanted to reform the calendar, they wanted to reform humanity.

These were precisely the concerns that emerged in the writings of mid-nineteenth century intellectuals. Often such writings cited earlier treatises almost verbatim, albeit presenting their claims as a reflection of the new spirit of modernization and enlightenment. In a proposal to reform the calendar dating from the first years of the Meiji period, Tsukamoto Aketake[12] harshly criticized the existing system, and, repeating the language of his predecessors, denounced the "ignorant fables that impede the advancement of people's knowledge."[13] An identical phrase is found in the imperial edict announcing the reform of 1873, which decries "ignorant fables that impede the advancement of human wisdom."[14]

The complexity of the luni-solar system was now framed as "inconvenience."[15] The compilation of a template for the printed calendar required the ingenuity of astronomers who found ways to reconcile the seemingly incompatible cycles of the moon, the sun, and the stars, and "granted the seasons" to the people. For reformers, however, the main problem with the calendrical template paradoxically lay in the fact that it was necessary. For hundreds of years, the calendrical template was seen as a pinnacle of human achievement that provided people with knowledge without which social conduct on a daily basis would be impossible. By the middle of the nineteenth century, however, the fact that people could not lead their lives without consulting the calendar came to be seen as an impediment. Calling for reform was neither a matter of improving the calendrical algorithm, nor of the sudden discovery of deficiencies in the previous calendar; rather, it was the association of specific features with broader cultural norms that made a luni-solar calendrical system inconvenient. For mid-nineteenth-century thinkers, the fact that astronomers were the only ones capable of creating order out of the complexity of celestial motion made the product of their work—the calendar—an obstacle to the proper management of society. The calendar was now seen as too indispensable to be based on the esoteric work of astronomers.

The reliance on other humans—skilled as they were—to complete arcane calculations came to be seen as an "inconvenience," while a system that was readily understandable came to be seen as desirable. In 1872, in a proposal to reform the calendar, Ichikawa Saigū[16] explained the benefits he believed the new calendar would bring:

> [It is] extremely convenient and extremely clear; a beautiful calendar that won't change in myriads of generations. In times of [extensive] international relations, it will show the world the unprecedented and uninterrupted imperial line [of Japan]; it will be enormously useful to navigation, and it will enable us to easily make a nautical almanac good for several years ahead; it will be possible to schedule with somebody an appointment that will happen in several decades, and know the eclipses that will happen hundreds of years from now. There is no inconvenient intercalary month, and once you get to know the number of days in a year there is no nuisance of remembering it each year.[17]

In the minds of thinkers like Saigu, the "enlightenment of the people" meant simplifying the calendrical system to the point that it was no longer necessary to rely on astronomer's yearly calculations at all. These calculations would be incorporated into the calendrical system itself. Anything that was not already built into the system, but relied on additional human intervention, came to be seen as an inconvenience.

The rhetoric of *inconvenience*, present in virtually every mid-nineteenth-century text that dealt with the (possible) reform, was framed as concern for "the people" who could not calculate dates on their own.[18] Ichikawa wrote that "our present system is based on the moon, and makes it easy to follow the phases of the moon, but for the distribution of months in one year it is very inconvenient." Under the new system that he proposes, however, "months are aligned with the seasons, which will be very useful to agriculture."[19] In an address to the parliament, another reformer stated that "because of the intercalary months, months do not match the *sekki* which is very inconvenient. There is a great damage to that. The people are used to this [system] and up until now they have not realized this enormous loss, but this day has come. We must change the calendrical system."[20] According to this logic, "the people" were "used to the system" and thus were too ignorant to realize its inconvenience, and it was the role of intellectuals to inform "the people" what was most convenient.

This rhetoric of enlightened usefulness, as opposed to ignorant inconvenience, culminates in the single most famous text about the Meiji calendrical reform—*The Treatise on the Calendrical Reform* (*Kairekiben*).[21] The treatise was written by the leading "modernizer" of Japan, Fukuzawa Yukichi, immediately after the new calendrical system was inaugurated.[22] Produced in a mere six hours one morning (according to the author), the several-page pamphlet became an instant hit. Fukuzawa Yukichi began by praising the simplicity of the new Meiji calendrical system:

> In this system, the four seasons are the same each year with no difference in heat or cold. Every day and every month falls exactly on the same day as the year before. And if you want to plant your seeds or harvest the grain, you don't need to look at the calendar—if in the previous year [the holiday] *higan* fell on the 21st day of the third month, then this year it will fall on exactly the same day.[23]

Later in the text he repeats this same point: a calendrical system based on lunar motion requires the consultation of an actual paper calendar to determine the seasons, but the Western system obviates written calendars altogether.[24]

Fukuzawa dismissed concerns by branding them as "ignorant." Similar to previous authors, this urban scholar seemed to know what was best for the agrarian population. He predicted that what he said "will at first appear to the unlearned people as surprising and inconvenient," but, he continued, "even though the inconvenience of ignorant people is pitiful, I don't have the time to explain everything to them."[25] Rather than arguing the validity of the new system, Fukuzawa preferred to shame people into accepting it. After all, nobody wanted to be associated with the "uneducated and ignorant fools" who dared to doubt the new system.[26]

In the spirit of the highly commercialized society of late-Tokugawa, early-Meiji Japan, Fukuzawa was also addressing the common concern with finances. Referring to the previous, luni-solar calendar, he described the discrepancy between the solar year and the twelve lunar months as "missing days" and compared it with lost money.[27] No matter that this discrepancy was corrected by the intercalary month (so nothing was really "lost"), and no matter that the new solar calendrical system, too, had to accommodate a similar discrepancy between the length of the solar year (365 days, 5 hours, 48 minutes and 52 second) and the number of days in the calendrical year, an equivalent case of "missing hours." Neither Fukuzawa nor his readers were interested in the actual calculations. The new rhetoric of time = money was powerful enough to convince those who still entertained doubt.[28]

In general, the new calendrical system was presented in an idealistic—and perhaps even naïve—way. Even though many of the reformers had studied some astronomy, none were professional astronomers and none were familiar with the actual practices of astronomical calculation. As such, they were not aware of the fact that a "perfect" calendar was virtually impossible. Or, perhaps, they just chose to ignore the irregularities in the new system they were proposing. Even though all the writers complained about the "inconvenient" intercalary month, none were troubled by the existence of a leap day in the new system, and many simply "forgot" to mention it. When talking about the eternally repeating calendar, the authors omitted the fact that in the new system months did not map consistently to the cycle of days of the week. Many writers expressed an idealistic expectation that the weather of particular days and months would remain consistent from year to year. But these details did not matter to the reformers; what mattered were the ideals of simplicity, usefulness, convenience, and enlightenment. They, of course, were not aware of the fact that at the same exact time there were people in Europe who raised similar questions about their *own* system, complaining that it was too complex and changed from year to year.[29]

Although for the most part the rhetoric of mid-century reformers closely echoed that of late eighteenth- and early nineteenth-century scholars, there was one exception—the call to reform the system of variable hours. For earlier thinkers, the variable hour system did not appear fall into the same category of "convoluted" calendrical factors that "impeded the moral development of society" as the luni-solar arrangement did. In fact, as we saw in chapter 2, clock-makers during the early nineteenth century devised new ways of building calendrical calculations into clock designs themselves. In this sense, the hour calculation system had already been simplified for humans by having been delegated to clocks. In spite of this mechanical solution, however, proposals to reform the calendar from the middle of the nineteenth century included calls to adopt the Western system of timekeeping and to abandon the variable hours in favor of twenty-four equal ones.

Surprisingly, none of the authors seemed to find it necessary to explain why this change was needed, nor to try to legitimize the change in the eyes of the public. Most of them just repeated what was already known from earlier manuals that explained how to read Western clocks, such as those by Fujimura Heizō and Ogawa Tomotada. Ichikawa Saigū simply stated the fact that "one day is divided into 24 hours, each of which is divided into 60 minutes, each of which is divided into 60 seconds."[30] The official government notice also simply stated the change:

> Up until now we divided the day into twelve hours according to the length of day and night. From now on we will rely on the hours of "jishingi", which divides the time of the day and night into 24 equal hours. From the hour of the rat until the hour of the horse are the 12 hours of "before horse", and the twelve hours after the [noon] hour of the horse will be called "after horse". . . . Up until now we referred to the hours by the character, but from now on we will do it by the number.[31]

The imperial announcement that declared the reform didn't even mention the hours at all. In a rare acknowledgment, Tsukamoto Aketake claimed in his proposal that the previous system of hour counting was "inconvenient for all sort of things," but never really explained what all these things were, shifting the conversation instead to an easy denunciations of the "middle section" of the calendar, which included the auspicious and inauspicious indicators used in divination.[32] Even Fukuzawa Yukichi suspended his venomous rhetoric in favor of straightforward explanation, repeating almost verbatim early nineteenth-century descriptions of the Western system with the addition of fashionable English transliterations:

> In the West, the day is divided into 24 hours. Their hour is equal to the "old" Japanese half an hour. This half-hour is divided into 60 units, each of which is called a "minute" (*miniuto*). This minute is further divided into 60 seconds (*sekando*). This second is generally the same as one movement of a pulse. Now, the face of the clock is divided into 12, the small hand making this circle twice. The long hand revolves around the clock 12 times. Noon and midnight are the origin of the 12 hours, and at this time both hands come to this place together. From there they are gradually moving right, and when the short hand points to 1 o'clock, the long hand made the whole circle of 60 minutes and came back to the place of 12 hours, and from there it goes on its next round. While the short hand moves half way from 1 to 2, the long hand travels half of the clock face for 30 minutes, and comes to the place of the 6th hour. Therefore, when looking at the clock, first look at the place that the short hand points to, and then look at the place where the long hand points. For example, if the short hand points to the place between 9 and 10, and the long hand points to the place of 2, it means that the time is 9 o'clock and 10 minutes. Now, when the short hand passes the middle point and gets closer to 10, and the long hand progresses to the point of 8, this means that the time is 20 minutes to 10. It means that it would take another 20 minutes for the long hand to come back to its origin at 12. In this way, you can know the 60 minutes from the clock face and know what is the hour and what is the minute.[33]

What motivated the reform of the hours? Why change a system that had worked fine and didn't have any apparent flaws? One reason to abandon a perfectly good system was the presence of Western watches, which had become ubiquitous. Ichikawa Saigū, for example, claimed that "lately everybody carries [an imported pocket watch] around and uses its divisions of an hour. It is a very trusty and useful device."[34] Other mid-nineteenth-century reformers shared this point of view, and even the language of the imperial edict, which ordered everyone to start using "the hours of the *jishingi*," suggested that the reform was about codifying a practice that already existed. However, we also must remember why European watches had become so popular in mid-nineteenth-century Japan in the first place. It was not simply a matter of availability, as availability can be viewed as a response to existing demand. It was not about the quality of watches either, since Western watches could be—and were, before the mid-nineteenth century—encased in Japanese dials to adapt them to the existing system. Neither was it about the ability of the Western pocket watch to answer emerging time-keeping needs—as we have seen throughout this book, both the existing Japanese clocks and temporal system were highly flexible and adaptable, and could be modified to accommodate changing daily needs. If anything, Western watches may have elevated the importance of the needs they answered, needs which otherwise might have remained inconsequential.

What changed was the context in which people situated their associations between Western pocket watches and Japanese clocks. The associations themselves were not new, yet by the middle of the nineteenth century they were no longer neutral but had become progressively more positive. Let us now look at this transformation of the network of associations with Western watches, as well as the new meanings these associations acquired during the beginning of the Meiji period.

But first, let us recap the transformations in associations with time measurement described earlier in the book. The gradual adoption of trigonometry and spatial geometry in Japanese astronomy, which replaced algebraic calculation, brought about a change in the way astronomers thought about and handled time measurement. Converting measured time into arcs and angles required smaller and smaller time units, while the measured time itself came to be perceived in spatial and geometrical terms. Furthermore, since astronomers were active users of their instruments, they constantly tinkered with timepieces, modifying them to fit both their emergent time measurement needs as well as their evolving conceptions of time measurement. By tinkering with clockwork they also created conceptual links between the movement of the mechanism—and especially of the pendulum—and the motion of celestial bodies.[35] When transposed into the field of geography to address questions about the shape of the earth, these conceptual links between timepieces, the time they measured, and the space that was framed by this time, acquired additional meanings. It was no longer space embedded in diagrams and projected into celestial motion; it was a concrete geographical space—tangible, walkable, and occupied by humans with distinct local characteristics. The pendulum clock that measured longitude thus became linked to maps and national space.[36] Then, these associations between timekeeping devices and different kinds of spaces were conceptually transplanted into musings about hypothetical navigation on the open sea. But the pendulum clock could not function on a rocking boat, and its obsolescence paved the way for the Western chronometer, which was not negatively affected by the marine environment. Yet all the conceptual links with the pendulum clock were now transferred to a Western timepiece. Furthermore, Western timepieces now came to be seen as related not only to national geography, but also to international travel and international relationships. Western watches were now perceived as something not only convenient to use for a very specific purpose, but also directly associated with global politics. And, owing to a linguistic confusion that resulted in a practical mix-up, all Western watches were treated as "chronometers," and hence mistakenly associated with global affairs.

It is thus not surprising that proposals to reform the calendar, and with it the hour-counting system, were paired with writings that focused on geography and navigation. In 1872 Takahashi Tamagusuku from Nishioji wrote both *On the Necessity of Reforming the Calendar* (*Kaireki subeki koto*) and *On the Necessity of Learning the Art of Navigation* (*Kōkai no justu wo manabashimu koto*). Then, in 1873, Hirokawa Seiken wrote *On the Reform of the Calendar* (*Rekihō kaikaku no koto*) paired with *On the Urgency of Geography* (*Chiri kyūmutaru koto*).[37] The motivation behind these pairings lay in the new meanings that the association between Western watches and the sciences played in the emerging Meiji discourse.

These associations of Western timekeeping with astronomy, geography, and navigation fed into the new ideology of the Meiji period. The early Meiji period was characterized by an intellectual movement that promoted learning and Westernization. Its Japanese name, *Bunmei Kaika* (文明開化), is usually translated into English simply as "Civilization and Enlightenment." However, these two words translate only the first two characters of the Japanese title. The other characters mean "opening," and can be translated as "opening to the West." The association of Western watches with astronomy, geography, and navigation answered all of the movement's aspirations.

According to Fukuzawa Yukichi, the dawn of Japanese modernity required every single person to master knowledge that was previously a prerogative only of the few. Fukuzawa was one of the most vocal promoters of the Meiji enlightenment, having authored numerous treatises about the sciences, most famously *On the Advancement of Learning* (*Gakumon no susume*) and *Illustrated Encyclopedia of Physical Phenomena* (*Kinmō kyūri zukai*).[38] The latter, especially, is often hailed as the first Japanese introductory text on the sciences.

Although Fukuzawa presented his focus on science as a novel aspiration that derived from the needs of the new Meiji regime, he in fact relied heavily on the writings of early nineteenth-century scholars. Fukuzawa, of course, did not admit that he relied on pre-Meiji knowledge. But looking at the contents of his books it becomes clear that Fukuzawa, who was trained in the Dutch Studies academy in his youth, based his writings on early nineteenth-century treatises on astronomy, surveying, and chemistry, sometimes reproducing texts and images almost verbatim. And, not unlike Honda Toshiaki half a century before him, Fukuzawa urged everybody to strive to become astronomers and geographers by gaining theoretical knowledge of those fields.

Western watches, which were already associated with astronomy and geography, were now seen as both the material manifestation of—and the means to—achieving the educational agenda set by mid-nineteenth-century enlightenment ideology. As we have previously seen, Western watches were perceived as devices that measured absolute cosmic phenomena and reflected a mathematically determined truth about the mechanism of the universe. This association now fed into the growing scientism of the mid-nineteenth century expressed by Fukuzawa. If astronomers and navigators measured time with Western watches, then owning and using a Western watch made one appear scholarly and learned. In this sense, Western watches were the perfect embodiment of both the personal and the national processes of becoming "civilized and enlightened."

And, of course, Western timepieces also appealed to the "opening" part of Civilization and Enlightenment. By definition, these devices were Western and thus signaled a movement toward European culture. They displayed Roman and Arabic numerals, and the units they measured were not just divisions and subdivisions of an hour similar to the Japanese *bu* and *byō*. They were, as Fukuzawa points out, *miniuto* and *sekando*—entities that were similar to the Japanese units of time yet had a very different feel about them. And it was not only that the watches looked and sounded Western. They were also associated with navigation, a development that allowed Japan to reach the shores of Western countries and to participate in the global political and economic community. Measuring time with Western watches was essential to the navigational practice of determining longitude. It also situated Japan on the global map of longitudinal lines, which, by the 1870s, had already been grounded in the Greenwich prime meridian. Western watches also allowed global communication about time. Consequently, Western watches, and thus the time they measured, came to be seen as symbols of global trade and politics. They were a material embodiment of recent advancements in Japanese astronomy, geography, and navigation, and the global aspirations associated with them.

As the associations with Western watches grew positive, the associations with existing Japanese mechanical timepieces became increasingly negative. These old clocks were now associated with the old ways the reformers were trying to eradicate. They announced time by pointing to animal signs and a nine-to-four count, symbols that had come to be exclusively associated with divinational practices. It is not that they were inherently divinational—the ancient meanings linking these symbols and numbers to the *Classic of Changes* (known to readers as *I-Ching*) had been forgotten by the Tokugawa period, to the extent that Tokugawa scholars had to dig around in old Chinese classics to recover them. Yet these symbols *were* sometimes used for hemerological practices popular during the Tokugawa period, which reinforced the association with calendrical divination. The old clocks also measured hours, the variations of which were indicated in the old calendars, decried by reformers for their despicable "middle section" of auspicious and inauspicious indicators. Consequently, even though neither the clocks themselves, nor the variable hours that they measured (and not even the animal signs that designated the hours), were necessarily or inherently related to divination, Japanese clocks were guilty by association.

Therefore, the evaluation of the old and the new systems of timekeeping was shaped neither by practical factors, nor by their inherent nature, but by associations these systems evoked. Even though the Western convention of measuring hours *in itself* was not necessarily more rational or even more commonsensical to the majority of population, it had become associated with "rational" and "enlightened" practices such as astronomy and navigation. And while the old system relied on calculations made by astronomers and was flexible enough to be adapted to changing social landscapes, it was associated with practices that reformers saw as backwards and damaging to society. Western watches were not exclusively used by astronomers, and did not benefit most people, but they were associated with astronomical and navigational practices. Old Japanese clocks, on the other hand, were not exclusively used by diviners, and were flexible enough to be tinkered with to adapt them to new needs, but they were associated with the old ways, and in this context tinkering came to be seen as an inconvenience.

Take for example a statement by Ichikawa Saigū that the "recently imported pocket watch is very convenient not only because it is portable—it is also very detailed and benefits social interaction."[39] The qualities Saigu cites were not unique to Western watches—portable Japanese timepieces could offer a similar level of detail and could certainly aid social interaction. The case for the apparent convenience or inconvenience of Western watches did not precede the assertion that the Western system was desirable. No, the case was made retroactively, after people had already formed an image of the Western system—and Western clocks—as beneficial and desirable. This image was not practically derived; instead, it was derived from a network of associations that connected the Western system to positive values.

The rhetoric of early Meiji reformers exposed their contradictory relationship with the past. Although their own perception of time measurement was a direct result of conceptual transformations of the eighteenth and early nineteenth centuries, they sought to emphasize the novelty of their ideas by denigrating the past. And so we witness a curious phenomenon—early Meiji reformers doing one thing while saying something completely different. As they replicated the practices of their predecessors, cultivated a similar web of associations as had existed ten years earlier, quoted previous scholars almost verbatim, and repeated the same explanations of the Western temporal system, they asserted a novelty that they claimed had emerged out of the Meiji reformation and the events that followed. Saigū had referred to Western pocket watches as "recently imported" even though, by the time of his writing, Western pocket watches had been imported to Japan for many decades. Writing in 1877, Kumagai Tōshū stated: "a clock is a machine that is associated with modernity. It helps people to know exactly whether the time is late or early and to not be embarrassed by being late to an appointment [. . .] This only happened four or five years ago."[40] And later, in his famous autobiography, Fukuzawa denigrated the timekeeping practices of the Edo period:

> In the last years of the Tokugawa regime, the Western fashion gradually permeated society, and all the high officials in the government began carrying imported gold watches in the folds of their dress. Those gentlemen must have paid two hundred *ryō* or even three hundred *ryō* for each of those watches, but very few of them knew how to read the time. A group of these august men would be talking together in the palace waiting room and one of them would ask, "What time would it be now? What does your honor's watch indicate?" The gentlemen next to him would take out a shining gold watch and would say, "It is already past one o'clock." The first Gentleman would say. "Is it, Sir?", looking at his own watch, "Where does the long hand of your watch point to, Sir?" "The long hand of my watch is about three o'clock, " the other gentleman answers. This conversation indicates that the time was about quarter past one. Such were the conversations between those ponderous personages *who did not know the real use of the watches*.[41]

Assuming that these authors were not intentionally trying to misrepresent the situation (and mislead future historians), we must conclude that they were finding new, significant meanings in already existing practices. The association with navigation was not new, but now navigation came to be seen as a rhetorical tool to articulate nationhood. Using a Western clock was no longer a question of personal convenience or inconvenience, but a matter of national ideology. Linking time measurement with a newly emergent sense of nationhood, Kumagai Tōshū claimed that the clock "benefits the whole nation (*kokka*)." Ichikawa Saigu, concerned with representing the Meiji nation to the outside world, claimed that that "in an era of international relations, [the reformed calendar] will show the world the unprecedented and uninterrupted imperial line [of Japan]." These new, powerful associations assigned timekeeping a degree of national importance built on existing conceptual and practical links between time measurement and cartography, with its suggestion of international engagement. The characterization of Western time as practical was thus determined a-priori, before it could be tested. And once it was decided that Western time was useful, the inconveniences and shortcomings of the Western systems were ignored, or perceived as easily fixable. It was "conveninet" because it was symbolic of the beginning of a new era. New timepieces counted a new kind of time, and signified a break from the past, echoing Fukuzawa's famous words "We have no history, our history begins today."[42]

Civilizing Time

According to the 1873 reform, Western clocks and watches were supposed to replace the Japanese timepieces that had been in use up until that point. But breaking from the past was not that easy. For one thing, the past was embedded in the material culture of time measurement, staring at the people from the faces of Japanese clocks. Japanese-style clocks were still everywhere, especially in the cities. Western watches had perhaps become increasingly available, but in the early 1870s they were still outnumbered by clocks made, or modified, in Japan. The most reasonable thing to have done would have been to just toss them away. And indeed, scores of Japanese clocks were simply discarded.

But tossing away an old clock was not easy, since clocks possessed intrinsic value that went beyond their function as timekeepers. First of all, there was their material value. Some clocks were luxurious, made out of expensive materials by famous masters. They were beautiful objects of art, too cherished to put in the garbage. The owners of these expensive clocks were the ones who could easily afford to replace them with newfangled Western ones, but they were still reluctant to toss away what was seen as an expensive toy. Such clocks were the most likely to survive and end up in museums or private collections, creating the impression that all Edo period clocks were objects of luxury. There were, of course, countless nonluxurious clocks, though those too were not cheap, and in a society that repaired even broken paper umbrellas the idea of disposing of relatively expensive devices was unacceptable. What's more, many clocks were family possessions and had a sentimental value, in addition to a monetary one.

The solution was therefore, again, to tinker with and to repurpose existing clocks. And so, the same clock-makers who had learned to modify Western watches and to affix Japanese dials to them now found a new job doing exactly the opposite—they positioned the movable digits at an equidistance and added dials with Arabic or Roman numerals to Japanese clocks, rendering them usable in the new era. This phenomenon perhaps started because it was cheaper and emotionally easier to "fix" old clocks than to buy new, expensive Western ones. But looking at the way these clocks were modified, we recognize that they came to serve a dual function.

The modified clock mapped the Western numerals onto the Japanese dial, turning clocks into time-translation devices. For those—unlike Fukuzawa and his fellow progressive intellectuals—who had not yet internalized the Western time system, these "fixed" clocks provided a material manifestation of the parallels between the two systems of measuring time (see color plate 10). Although time was measured in units that did not possess intrinsic meaning, people associated certain digits with the activities often done during those particular hours, such as working, dining, or playing, or with phenomena that occurred during those times such as noon or midnight. One of the difficulties Japanese users faced was the fact that in the Western temporal system there were no hour digits associated with dawn and dusk. During the last several decades of the Edo period, as well as the first years of the Meiji period, Japanese users relied on written manuals, which instructed them—visually and numerically—how to translate their experience of time measurement into the Western system and vice versa. But these manuals, even those with abundant images that represented the changes of time during various seasons, were static—one had to chose a particular moment to translate between systems. The modified clocks of early Meiji, however, allowed their owners to continuously follow the processes of time measurement while comparing the two systems. By following the movement of the clock according to both systems, early Meiji clock users were able to experience both systems simultaneously and to connect the alien temporal units to the lived experiences of time measurement that had been previously associated with the animal signs of the variable hours. With these modified clocks, Meiji users could gradually adapt their new temporal practices to their previous habits and associations with time measurement.

And there was indeed a need for time to internalize and adjust to the new system.[43] To the great dismay of reformers, the general populace was not quick to break with the past and embrace the new system. Many—"uncivilized" in the eyes of reformers—still insisted on engaging in divinational practices and seeking out auspicious and inauspicious days and directions. The government tried unsuccessfully to ban these practices, which only drove the market for divinational calendars underground, divorcing it from astronomers' calculations and from astronomy in general. Divination practices did not disappear even in the twentieth century, but only changed to include Western style horoscopes.

Even for those who sincerely did want to reform their measurement of time, the task was not easy, as they needed to overcome a set of existing assumptions rooted in associations with the previous system. Changes in the material environment—the manuals, the Westernizing of timepieces, the establishment of public clocks—all helped. But even with these changes, as well as the instruction in the new mode of timekeeping provided by various social and educational institutions, the transition was slow. Unlike the intellectuals who linked the new system of timekeeping to a variety of scientific, philosophical, and ideological developments, many people on the streets of what was now called Tokyo thought about timekeeping in a way that remained deeply entrenched in the old networks of associations.[44]

Despite the difficulties and the slow process of accustomization to the new system, the reform did eventually succeed and Western conventions of timekeeping took root in Japan. More and more people carried Western watches, large Western-style dials were installed in public places, and the newly emergent train schedules were written according to the system featuring twenty-four hours in a day, sixty minutes in an hour, and sixty seconds in a minute. The old way of reckoning time seems to have gradually receded from memory.

A change from one way of dividing the day to another, however, does not neatly translate into changed perceptions of time or timekeeping. The conventional division of the day into particular units serves as a prerequisite for the development of timekeeping practices, temporal habits, and perceptions of time and its measurement. Yet there is a limit to the extent to which temporal conventions can determine perceptions of time. The two are intertwined and mutually dependent, yet distinct.

Consequently, specific timekeeping practices cannot be explained by a timekeeping system alone. When discussing the success of the Meiji reform of the temporal system, people often point out the fact that Japan now boasts what might be seen as the most precise time-measurement practices on earth. Yet this narrow sense of precision, which boils down to seconds on train schedules, does not derive from the system of 24 hours per se. After all, in most of Europe, and certainly in the United States, the precision of train schedules does not even approach that of Japan. The convention of counting time using twenty-four hours divisible into sixty minutes divisible into sixty seconds only provides a framework within which the scheduling of trains takes place. The origin of specific timekeeping practices lies elsewhere—in the activities that forge associations that define the meaning of the seemingly neutral act of measuring time.

Conclusions

Conceptions of time develop over time and with respect to time-related practices. They may seem stable in a dictionary-style entry, which supposedly reflects what is perceived to be the best definition that captures the nature of time. Yet if we look at conceptions of time *in use*, we discover that they are anything but stable and well-defined. The meanings people attribute to time and its measurement are significantly conditioned by their own associations with lived experience and the particular contexts in which those associations are formed. These associations are formed in the course of everyday practices by which people invest tools with the power to measure and interpret abstract phenomena. How, for example, is the progression of time represented on our timekeepers? What is time measurement needed for? Where do we look in order to find the "now"? The varying answers to these questions form associations, the aggregate of which constitutes our concept of time and its measurement.

Consequently, conceptions of time *in use* have history. Being rooted in an endless stream of associations with concrete conventions, bodily habits, calculatory practices, images, etc., conceptions of time, *held by particular humans*, are in a constant flux, varying among groups of individuals and always changing. A history of a particular conception of time is therefore not a history of evolution of dictionary definitions, but rather a history that reflects the processes in which conceptions of time develop in the minds of people on the basis of the time-related practices with which they are familiar.

Focusing on the development of the time-measurement practices of Tokugawa Japanese astronomers, this book has described one such history. It has described how conceptions of time arose out of pendulum swings, angle calculations, seasonal terms, graphs on paper sundials, sextants, time-bells, Chinese astronomical treatises, clock-makers' tools, trigonometrical calculations, mechanical models, pictorial representations of seasonal changes in the lengths of hours, movements of index hands on clock dials, longitude measurements, castaways' stories, timekeeping conventions, habits, etc.

I have described the processes by which Tokugawa astronomers incorporated into their practices more and more fragments of European culture that aligned with European-style timekeeping conventions—processes that resulted in changes to their normative evaluation of European timekeeping from "nonsensical" (in the early Tokugawa period), to "comprehensible" (in the eighteenth century), to "desirable" (by the middle of the nineteenth century). European timekeeping conventions gradually made more sense *in relation to other practices*. With European practices of drawing diagrams as a method of trigonometrical calculation, counting the oscillations of pendulums, and making observations using sextants, it made sense to conceive of time in terms of celestial arcs. When applied to geodesy, translated to maps, and utilized for the calculation of longitude, the arcs of time could be seen as part of a single belt of time wrapping around the earth. Since arcs of time were defined by celestial motion, the belt of time around the earth was, too, conceived in terms of the celestial motion of the earth around the sun. And since European atlases depicted the earth as divided into twenty-four temporal segments, and British nautical almanacs treated the time equation in European temporal units, the whole system of twenty-four equal hours now made sense, at least for those people who dealt with questions of astronomy, geography, and navigation.

I have also described the processes by which astronomical time came to be seen as primary, as the only time that mattered. Tolerance for multiple coexisting conceptions of time and a plurality of time-measurement methods gradually gave way to the idea that all measurements of time, regardless of the purpose or the context, should accord with astronomical practice. Up until the nineteenth century, the timekeeping system used by astronomers was just one of the possible ways to measure time. Even astronomers themselves did not claim that the system they used at work was in any way superior to the timekeeping system they used for everyday purposes—scheduling a meeting, or visiting a governmental office or temple. Moreover, as we saw in chapter 3, astronomers, too, saw timekeeping systems as malleable as they juggled, modified, invented, and adopted a variety of temporal units and timepieces to fit their particular observational and calculational needs. However, as the idea of time as an independent mathematical entity took root, and as astronomers came to see Western timepieces as better suited to measure mathematical Time, their tolerance of the multiplicity of time-measurement systems began to wane. At the same time, the discourse that started with Honda Toshiaki, and which positioned astronomy, geography, and navigation as activities essential to proper governance, contributed to an increasing desire to follow astronomical practices—particularly the ones that used mathematical calculations and precision instruments. By the mid-nineteenth century, as astronomers settled on Western chronometers as their preferred technology of time measurement, the Western chronometer's timekeeping system became no longer just one possible way to measure time—it became *the* way to measure Time.

By the beginning of the Meiji period, the Western timekeeping system and Western timepieces came to be associated with numerous positive values—old and new. Although late Tokugawa and Meiji thinkers often justified their policy proposals in terms of "convenience" and "benefit to the nation," the act of labeling something as "convenient" or "inconvenient" was, in itself, rooted in wider associations with favorable versus unfavorable norms. For most of the Tokugawa period, people found ways to tweak their devices, modify their practices, and even adjust their own bodies to overcome practical troubles and escape labeling their timekeeping system as "inconvenient." Thus, for example, during the period when Japanese astronomers hailed the pendulum clock as the pinnacle of technological achievement, they did not find it "inconvenient" to avoid breathing heavily near the pendulum or to account for the daily rate of oscillation during cold weather. On the other hand, once the Western time system became associated with scientific and social progress, progressive thinkers preferred to overlook the fact that the Western system, too, had its problems, and that Western clocks, too, tended to run slow or late. The labeling of the existing system of variable hours as "inconvenient" had more to do with its association with prognostication practices than with its alleged incompatibility with modern life. And similarly, the praising of the Western temporal convention as "convenient" was due to its association with astronomy, mathematics, international commerce and politics, progress, enlightenment, and above all—the West itself.

In a sense, it is possible to say that the newly emergent conception of time in late Tokugawa Japan was "modern." Yet it was not "modern" in the qualitative terms that have classically been associated with modern time—as a degree of abstraction, in relation to modes of production, or in its use in an urban setting. No, the emerging astronomical notion of mathematical time was as rooted in concrete practices (albeit those of astronomers) as other conceptions were, and was not actually necessary for the time-management of rapidly industrializing Japanese mega-cities. It was modern, however, simply because it was the prevalent conception of time during the historical period we typically label "modernity."

Beyond the history of transformations in Japanese perceptions of time measurement, however, I have argued that humans interpret new bits of knowledge by associating them with elements that are already familiar to them. Whether new knowledge comes in the form of a device, an image, a technique, a text, or a spoken word, it is inevitably interpreted through association with already existing practices. Furthermore, people's actual use of foreign devices or techniques too was only possible once those were conceptually situated in a landscape of familiar material environment.

This view of technological and scientific practice as embedded in concrete associations bears several methodological implications. First, it means that our interpretation of technology is driven not only by practical needs, social norms, and judgments, but also by the images, memories, and bodily motions we associate with similar technologies we encountered in the past. As we have seen, mechanical timekeeping technologies were interpreted in Japan through associations with other, nonmechanical timepieces. Clock-makers came up with new designs based on their knowledge of how time behaves on incense clocks, or in clepsydra, and how the yearly changes were represented in graphs and tables. All Tokugawa clock-makers attempted to design clocks that would fulfill their social function—to measure hours of variable length. But the ways they did so differed according to the shifting relations between practices, images, and things that they knew were related to timekeeping.

A second implication of this observation is that the spread of technology and knowledge depends on the spread of common associations. If the interpretation and implementation of a tool depends on the web of associations in which it is given meaning, then acquiring a shared set of associations is necessary for technology transfer. Without associations rooted in European cultural experience, Japanese clock-makers and users interpreted mechanical timekeeping technology in ways strikingly different from how it had been interpreted in Europe. Objects indeed left the shores of Lisbon or Amsterdam and arrived in Nagasaki, but does this fact alone mean that we have witnessed a transfer of technology? Or is it only when the local interpretation resembles the one at the place of origin that we can say that a technology (and not merely an object) has been transferred? What degree of similarity of practice and purpose is needed to achieve the transfer of technology?

It was only when Tokugawa astronomers' associations with time measurement came to include European elements that their conceptualization of time measurement came to resemble the one in Europe. It was the new association of time measurement with trigonometry, diagrams, arcs and angles, and sextants and pendulums that paved the way to an understanding of time in a manner similar to (but still not quite the same as) that of European astronomers.

This observation applies to the transfer of technologies within Japan as well. Even at the close of the Tokugawa era, most Japanese did not share their astronomers' understanding of time measurement. The understanding of time measurement as mathematically related to geographic locations was initially limited to scholars like Honda Toshiaki, officials like Kaga *daimyō* Maeda Norinaga, clock-makers like Tanaka Hisashige, and students of Western science like Fukuzawa Yukichi. For others to understand how Western timekeeping worked, more was needed than the theoretical outlines of the foreign system. Such explanations had to include the Western assumption that the count of hours begins at the top of the dial, or that Western clocks do not indicate dawn and dusk, etc. Ordinary citizens in the late Tokugawa and early Meiji periods—who, unlike Fukuzawa Yukichi, had not received extensive education in Western science—had to first experience Western time in relation to the familiar conception of time they encountered in Japanese tables or modified clocks. Only then, once they had forged their own lived associations with this new system, could they gain a better understanding of Western timekeeping system.

The third implication of seeing technological practice as shaped by individual associations involves shifting our focus from direct, verbal transmission of knowledge to a consideration of the broader ecology of technological practice. The particular political situation in Tokugawa Japan created the rare situation in which nearly all of the very minimal contact with foreigners was documented. We also know that the linguistic abilities of interpreters only reached the level necessary for accurate translation in the nineteenth century—after Japanese astronomers already expressed ideas similar to their European colleagues. Thus, in the narrative that has unfolded in this book, we can know for sure that Japanese astronomers did not just repeat what they had learned from the Westerners. They formed those ideas because they were dealing with similar problems, using similar methods, and relying on similar associations to guide their conclusions. They found implicit meanings in pictures, in foreign maps, in diagrams, and in instruments. And they came to conclusions that the universe behaved similarly to their timepieces, that time measurement bound geographical places with stars, and that time unified nations. Eventually, as we have seen, this led them to accept the proposition that Western time was a more accurate reflection of astronomical reality.

All of these points suggest that historians, sociologists, anthropologists, and other social scientists interested in the social shaping of technology should revisit the kind of social factors that form human interaction with technology. Political events, social norms, economic considerations—these all undoubtedly influence the ways humans deal with technology. But although these dimensions of social conduct may explain some of the motivations behind certain practices, they cannot explain the particular forms those practices would take. Thus, for example, they can explain the fact that Japanese artisans sought to adapt Western mechanisms to measure hours of variable length, but they cannot explain how artisans devised the particular designs they ended up creating. Social categories, alone, cannot explain why there were not one, but rather several clock designs, all of which were used contemporaneously. So, too, are these categories a bit too broad to explain the mapping project of Inō Tadataka, which originated in an obscure astronomical problem that Takahashi Yoshitoki happened to fixate on. Nor are they sufficient to explain how Hazama Shigetomi came to imagine the universe as a set of pendulums hanging from the sun. All of these individuals came up with their ideas following a distinct set of encounters, interactions, and experiences—refracted, of course, through the different social contexts which they lived through—but formed on a much more minute level than those invoked by phrases like "economic reforms" or "political pressures."

* * *

Writing this book required confronting the irony of narrating a history that brought about the denial of the historicity of the concept of time. In it, I have described a process by which particular historical circumstances resulted in the emergence of a new concept of Time, with a capital T, that does not possess either history or context. Though rooted in astronomers' concrete methods of measurement, calculation, and representation, this newly emergent conception of Time dictated that it existed independently of human activities, in celestial motion and mathematical algorithms.

Yet this seemingly ahistorical Time is just a façade. The transition to the new conventions of dividing the day did not result in a complete overhaul of the existing network of associations. In fact, traces of old temporal conventions continued to persist in the dark corners of temporal practice. The practices, habits, and associations that contradicted the new ideological aspirations of the Meiji period were weeded out and eliminated. But there were numerous others that did not necessarily conflict with the emergent temporal regime. Today, if we look closely, we can recognize attitudes that have survived the centuries of social, ideological, and material changes. One of these can be seen in the general acceptance of the fact that the system of timekeeping is malleable and tweakable. Walking around big Japanese cities today, it is not unusual to spot signs on the doors of restaurants or bars announcing that the place is open until 25, 27, or even 28 o'clock. There is no doubt that in today's Japan there are no people who do not accept the division of the day into twenty-four equal hours. Yet it appears that the longstanding tacit assumption that there are different "times" that require different approaches, some of which can be modified to fit a specific time-measurement purpose, survived the temporal revolutions of the late nineteenth and the twentieth centuries. Of course, there are twenty-four hours in a day, but in twenty-first-century Japanese night-life time, it is perfectly rational to say that the bar is open until 28:00 in the morning. If you plan on staying up late, you will understand.

ACKNOWLEDGMENTS

While researching and writing this book I benefited from numerous individuals who shaped my thinking, encouraged me during difficult times, and offered practical support. I am deeply grateful to all the colleagues, teachers, mentors, editors, administrators, students, friends, and family members who supported the process in various ways. Although I could not possibly thank everybody by name, I am deeply grateful to each and every one of them.

In support of this work I received generous funding from the Japanese Government Ministry of Science and Education (Monbukagakushō Scholarship), the Cressant Foundation, the Japan Foundation, and the Princeton Institute for International and Regional Studies. This funding enabled me to conduct archival research in Japan to develop the abilities necesassary to engage with modern and early modern documents.

The course of my study took me through various institutions in different parts of the globe. The intellectual foundations for this project were laid at the Cohn Institute for the History of Sciences and Ideas and the East Asian Studies Department at Tel Aviv University. Princeton colleagues and mentors had a most profound role in shaping this study. I am particular grateful to the late Mike Mahoney, who taught me to pay attention to what historical actors actually did (and not only to what they said); to Ben Elman, who insisted that if historical reality did not match my initial hypothesis it only meant that actual history was more interesting than I had previously imagined; and to Michael Gordin, who could always suggest three possible directions my research might take. In Japan, I have continually benefited from the kindness of Professor Takehiko Hashimoto, whose research on Japanese clocks helped me with my own research on astronomical time measurement. Professor Suzuki Kazuyoshi of the Tokyo National

Science Museum, and Professors Nakamura Tsukō, Yoshida Tadashi, Mitsuru Sōma, and Motoyasu Hiroshi all shared their invaluable expertise in early modern Japanese science with me. The Seiko Museum kindly gave me access to their collection of Tokugawa period devices. I am also forever in debt to all the wonderful teachers at the Inter-University Center for Japanese Studies in Yokohama. More recently I have enjoyed stimulating conversations at the Max Planck Institute for the History of Science, and benefited from the advice and support of my friends and colleagues at the Johns Hopkins University.

I am particularly grateful to my friends and family for their ongoing help and support. Their love and care accompanied me throughout the years of research and writing. I could probably make a book-long list of friends who offered advice, inspiration, and moral support. Above all I am thankful to Loren Ludwig, who by now probably knows this book by heart.

APPENDIX 1: HOURS

Each hour in Tokugawa Japan was associated with a particular number, character, and meaning expressed in the conventional reading of characters as animals. For mnemonic purposes, the sequence is recited in four series of three, starting from *Ne*.

9	子	*Ne*	Rat	Midnight
8	丑	*Ushi*	Ox	
7	寅	*Tora*	Tiger	
6	卯	*U*	Rabbit	Beginning of the day
5	辰	*Tatsu*	Dragon	
4	巳	*Mi*	Snake	
9	午	*Uma*	Horse	Noon
8	未	*Hitsuji*	Sheep	
7	申	*Saru*	Monkey	
6	酉	*Tori*	Rooster	End of the day
5	戌	*Inu*	Dog	
4	亥	*I*	Wild boar	

APPENDIX 2: SEASONS (*SEKKI*)

立春	*Risshun*	Beginning of Spring
		Indicates the first month of the year
雨水	*Usui*	Rain Water
啓蟄	*Keichitsu*	Awakening from Hibernation
春分	*Shunbun*	Spring Equinox
清明	*Seimei*	Clear and Bright
穀雨	*Koku-u*	Rain on the Grain
立夏	*Rikka*	Beginning of Summer
小満	*Shōman*	Gradual Ripening
		Literally: Things get a little fuller
芒種	*Bōshu*	Grass and Grain
夏至	*Geshi*	Summer Solstice
小暑	*Shōsho*	Minor Heat
大暑	*Taisho*	Major Heat
立秋	*Risshu*	Beginning of Autumn
処暑	*Shosho*	Final Heat
白露	*Hakuro*	White Dew
秋分	*Shūbun*	Autumn Equinox
寒露	*Kanro*	Cold Dew
霜降	*Sōkō*	Frost Descending
立冬	*Rittō*	Beginning of Winter
小雪	*Shōsetsu*	Minor Snow
大雪	*Taisetsu*	Major Snow
冬至	*Tōji*	Winter Solstice
小寒	*Shōkan*	Minor Cold
大寒	*Taikan*	Major Cold

NOTE. Translations of *sekki* names in the third column are author's own.

APPENDIX 3: YEARS IN THE *NENGŌ* SYSTEM

In Japan, years are conventionally referred to as year X in era Y, or as the *e-to* (*kanshi*) of era Y (see appendix 4). The third column below shows the *e-to* (*kanshi*) of the first year of a particular era. The last column gives the first year of a particular era according to the Gregorian calendar.

元和	*Genna*	乙卯	1615
寛永	*Kan'ei*	甲子	1624
正保	*Shōhō*	甲申	1644
慶安	*Kei'an*	戊子	1648
承応	*Jōō*	壬辰	1652
明暦	*Meireki*	乙未	1655
万治	*Manji*	戊戌	1658
寛文	*Kanbun*	辛丑	1661
延宝	*Enbō*	癸丑	1673
天和	*Tenna*	辛酉	1681
貞享	*Jōkyō*	甲子	1684
元禄	*Genroku*	戊辰	1688
宝永	*Hōei*	甲申	1704
正徳	*Shōtoku*	辛卯	1711
享保	*Kyōhō*	丙申	1716
元文	*Genbun*	丙辰	1736
寛保	*Kanbō*	辛酉	1741
延享	*Enkyō*	甲子	1744
寛延	*Kan'en*	戊辰	1748
宝暦	*Hōreki*	辛未	1751

明和	*Meiwa*	甲申	1764
安永	*An'ei*	壬辰	1772
天明	*Tenmei*	辛丑	1781
寛政	*Kansei*	己酉	1789
享和	*Kyōwa*	辛酉	1801
文化	*Bunka*	甲子	1804
文政	*Bunsei*	戊寅	1818
天保	*Tenpō*	庚寅	1830
弘化	*Kōka*	甲辰	1844
嘉永	*Ka'ei*	戊申	1848
安政	*Ansei*	甲寅	1854
万延	*Manjū*	庚申	1860
文久	*Bunkyū*	辛酉	1861
元治	*Genji*	甲子	1864
慶応	*Kei'ō*	乙丑	1865
明治	*Meiji*	戊辰	1868
大正	*Taishō*	壬子	1912
昭和	*Shōwa*	丙寅	1926
平成	*Heisei*	己巳	1989
?			20??

APPENDIX 4: THE *KANSHI*, OR *E-TO*, CYCLE

The sexagenarian—*kanshi* or *e-to*—cycle is generated by combining the ten "stems" (*kan* 干) with the twelve "branches" (*shi* 支). The ten *kan* are 甲 *kō*, 乙 *otsu*, 丙 *hei*,丁 *tei*, 戊 *bo*, 己 *ki*, 庚 *kō*, 辛 *shin*, 壬 *jin*, 癸 *ki*. The twelve *shi* are the same animal signs that are used for the hours (appendix 1), but pronounced *shi, chū, in, bō, shin, shi, go, bi, shin, yū, jū, gai*. In constructing the cycle, the first *kan* is combined with the first *shi*, the second *kan* with the second *shi*, etc. (After the tenth kan, the first kan is combined with the eleventh shi, etc.) The ten "stems" also correspond to the Five Phases—Wood, Fire, Earth, Metal, Water—and divided into "elder brother" (*e*) and "younger brother" (*to*), giving the system its alternate name, *e-to*, written using an alphabet rather than characters.

The sequences of *kan* and *shi* were also commonly used as counters—similar to how people in the West use a,b,c to indicate 1,2,3. Thus, for example, astronomers used *kan* and *shi* to label specific points on diagrams, and when making calculations could refer to an angle as 甲乙丙.

甲子	乙丑	丙寅	丁卯	戊辰	己巳	庚午	辛未	壬申	癸酉
1	2	3	4	5	6	7	8	9	10
甲戌	乙亥	丙子	丁丑	戊寅	己卯	庚辰	辛巳	壬午	癸未
11	12	13	14	15	16	17	18	19	20
甲申	乙酉	丙戌	丁亥	戊子	己丑	庚寅	辛卯	壬辰	癸巳
21	22	23	24	25	26	27	28	29	30
甲午	乙未	丙申	丁酉	戊戌	己亥	庚子	辛丑	壬寅	癸卯
31	32	33	34	35	36	37	38	39	40
甲辰	乙巳	丙午	丁未	戊申	己酉	庚戌	辛亥	壬子	癸丑
41	42	43	44	45	46	47	48	49	50
甲寅	乙卯	丙辰	丁巳	戊午	己未	庚申	辛酉	壬戌	癸亥
51	52	53	54	55	56	57	58	59	60

NOTES

INTRODUCTION

1. Robertson, *The Evolution of Clockwork*, p. 194.
2. While there is no debate that the Tokugawa period ended in 1868, historians still disagree about when it began—1600 is as close to a consensus choice as any date. As for nomenclature, Tokugawa is the name of the shogunal family that ruled during the period, and Edo is the name of the capital city, which later became Tokyo. "Tokugawa period" is often used in discussions of political affairs, while "Edo period" is often favored when talking about culture.
3. Robertson, *The Evolution of Clockwork*, p. 191.
4. In the words of Japanese authors, equal hours "did not distinguish between long and short hours." See for example Nishikawa Joken, cited in Hirai, *Tokei no hanashi*, p. 129.
5. E. P. Thompson, "Time, Work-Discipline and Industrial Capitalism," p. 60. Thompson's thesis was widely repeated in the 1980s and 1990s, in studies such as Landes' *Revolution in Time*, Turner's *Of Time and Measurement*, and Postone's *Time, Labor, and Social Domination*. In Japanese studies, Thompson's influence is apparent in Tanaka's *New Times in Modern Japan*, Shimada's "Social Time and Modernity in Japan," and Asao's "*Jidai kubunron*." Thompson's thesis is still popular today, despite a series of studies that show how the transformation of temporal order in Europe and America had significant influences besides industrialization. See Dohrn-van Rossum's *History of the Hour*, Sauter's "Clockwatchers and Stargazers," and McCrossen's *Marking Modern Times*. For a recent discussion of how modern technologies brought about the opposite of modern time discipline in Egypt, see Barak's *On Time*.
6. On the transformations in temporal order during the Meiji period see especially Hashimoto and Kuriyama, eds., *Chikoku no Tanjō*, and Nishimoto's *Jikan ishiki no kindai*.
7. The most famous of whom is Fukuzawa Yukichi (1835–1901). This claim features in much of his writing, most notably in *Kairekiben*.
8. Which holds true not only for mid-nineteenth century Japan but also for the highly industrialized landscape of early twentieth-century England, as shown by Rooney and Nye in "'Greenwich Observatory Time for the Public Benefit.'"
9. McCrossen makes this point clear in her *Marking Modern Times*.
10. Edgerton, *Shock of the Old*.

11. Smith, *Native Sources of Japanese Industrialization,1750–1920*, pp. 203–19, and Ministry of Justice Research Dept., *Tokugawa minji kanreishū*, vol. 5, pp. 113–15.
12. The idea that changes in temporal practices were provoked by trains and telegraphs had already been debunked by scholarship examining European cases. See Rooney and Nye, "'Greenwich Observatory Time for the Public Benefit.'"
13. I have explored how Tokugawa Japanese viewed the European time system and clocks in "Translating Time: Habits of Western Style Timekeeping in Late Tokugawa Japan."
14. Although there were pockets of Europe where variable hours were used well into the nineteenth century. Birth, *Objects of Time*, p. 49.
15. For a discussion of this proto-capitalist economy, see Howell's *Capitalism from Within*.
16. Adas's *Machines as the Measure of Men* and Winichakul's *Siam Mapped*. See also the recent volume edited by Sandra Harding, *Postcolonial Science and Technology Studies Reader*. Specifically, on the topic of power struggles accompanying the diffusion of the Western notion of time, see Nanni, *The Colonisation of Time*. Barak focuses on resistant and subversive uses of technology in *On Time*.
17. Rebellions known as *ikkō-ikki*, driven by adherents of the True Pure Land branch of Japanese Buddhism, had taken place between the mid-fifteenth and late sixteenth centuries.
18. Foreigners—Portuguese, English, and Dutch—originally had a post in Hirado, at the westernmost point of the southern island of Kyūshū. But even that remote spot was not far enough away, and in the 1640s, the trading post was moved further south to the newly constructed Dejima, effectively confining the foreigners to an island.
19. Chaiklin, *Cultural Commerce and Dutch Commercial Culture*.
20. 南蛮 and 紅毛.
21. According to Shunkai's student Tani Shinzan. Cited in Nakayama, *Nihon no Tenmongaku*, p. 75.
22. In his *Matters of Exchange*, Harold Cook states that both the Dutch and the Japanese were interested in "things that work." The problem is that certain things were often deemed to be "working" only to be declared useless later, and vice versa. The question, then, is what made people decide whether something "worked" or not.
23. Beukers, "Dodonaeus in Japanese: Deshima Surgeons as Mediators," p. 285.
24. I have explored developments in scientific translation in Japan circa 1800 in "Before Words: Reading Western Astronomical Texts in Early 19th Century Japan."
25. Shirahata, "The Development of Japanese Botanical Interest and Dodonaeus' Role," pp. 267–68.
26. Historians of science have discussed the influence of astronomical practices on a wider range of social, cultural, and political activities. See particularly an edited volume on the role of nineteenth-century observatory sciences in Aubin et al., *The Heavens on Earth*. In *A Tenth of a Second*, Canales describes how the observatory-born notion of "personal equation" spread through a series of unrelated fields of study.
27. Ferguson, *Engineering and the Mind's Eye*.
28. Fukuoka, *The Premise of Fidelity*.
29. Marcon, *The Knowledge of Nature and the Nature of Knowledge in Early Modern Japan*.
30. On the famous debate between Einstein and Bergson over the nature of time, for example, see Canales, *The Physicist and the Philosopher*.

31. In *A Tenth of a Second* Canales has explored one such process—how the notion of a "tenth of a second" crystallized into an unquestionable paradigm.
32. Andrew Pickering called this array the "mangle" of practice.
33. In the context of science and technology in East Asia, particularly noteworthy is Dagmar Schäfer's *The Crafting of the 10,000 Things*, which shows how engagement with crafts shaped interpretations of metaphysical theories in late Ming China.
34. In *Einstein's Clocks, Poincaré's Maps*, Peter Galison discussed how Einstein's and Poincaré's conceptualization of time was rooted in their respective practices with concrete objects. Pickering mentions associations as one important element in the mangle of practice (*The Mangle of Practice*, p. 95). Here, I expand on this notion.
35. E. P. Thompson, "Time, Work-Discipline and Industrial Capitalism"; Landes, *Revolution in Time*; Turner, *Of Time and Measurement*, and Postone, *Time, Labor, and Social Domination*.
36. In history of technology, the approach known as SCOT (social construction of technology). For seminal examples of SCOT, see Bijker et al., eds., *The Social Construction of Technological Systems*, and especially Bijker and Pinch's "The Social Construction of Facts and Artifacts."
37. Smith, *Native Sources of Japanese Industrialization,1750–1920*; Morris-Suzuki, *The Technological Transformation of Japan*.
38. In sociological analysis, conventions are treated as both reinforcing and reinforced social structures. Bordieu, *Outline of a Theory of Practice*.
39. Latour, "Do Scientific Objects Have a History? Pasteur and Whitehead in a Bath of Lactic Acid."
40. Daston, "On Scientific Observation," p. 103. See also *Things That Talk* and *Biographies of Scientific Objects*.
41. Norton Wise wrote that "a machine functioning within a social context carries with it, simultaneously, a set of ideas (as both concepts and values), which explain its physical operation and the set of ideas that explain its social function. . . . I therefore say that it 'embeds' our ideas." "Mediating Machines," p. 79.
42. See Michael S. Mahoney's unpublished paper titled "Reading a Machine."
43. Polanyi, *Personal Knowledge*. Although Polanyi's original book was published in 1958, the notion of "tacit knowledge" is continually discussed by historians of science and technology. See, for example, works by Harry M. Collins.
44. Stephen Turner decribed it as "contrast" or "alternative." *Understanding the Tacit*, pp. 25–26.
45. Baird, *Thing Knowledge*.
46. See Aubin et al., *The Heavens on Earth*.
47. Galison described the process of the distancing from the Newtonian notion of absolute time during the late nineteenth and early twentieth centuries in Europe. *Einstein's Clocks, Poincaré's Maps*.
48. See, for example, Wood, *The Power of Maps*.
49. For many years, the Tokugawa period was characterized in historiography as confined by its "seclusion policy," or by the Japanese term *sakoku* (鎖国), which can be translated as "locked country." The latter term was actually coined (in German: *Landesabschließung*) by Engelbert Kaempfer, who visited Japan between 1690 and 1692, and wrote a multivolume study, *The History of Japan*. The Japanese word came into existence only when extracts of Kaempfer's work were translated into Japanese in 1801 by Shizuki Tadao. In the past few decades, historians have shown that

Tokugawa Japan was far from secluded, and maintained both commercial and intellectual ties with the outside world.

50. Ogle describes the spread of the notion of "global time" as the emergence of a new ideology rather than particular practical adaptations. *The Global Transformation of Time.*

CHAPTER ONE

1. For most of the Tokugawa period 時 (*toki/ji*), 刻 (*koku*), 剋 (*toki/koku*), and 辰 (*toki/koku/shin*) were used interchangeably. A separate system was sometimes used for the nighttime hours, in which they were referred to as term *kō* (更). When this system was implemented the night was divided into five equal *kō* and each *kō* was divided into five equal *ten* (点). Bell keepers would have a special conversion table that showed them at what hour (時) the beginning of each *kō* should be marked. Minamoto, *Tenji Meikai*, p. 25.
2. Honda, *Shisei chiji kagami*. If we translate the figures Honda Toshiaki calculated into modern terms, they come to more precisely to 76.75 and 155.952 minutes, respectively.
3. 不定時法.
4. And equivalents in other European languages.
5. *Shi* 支. Those were combined with ten "celestial stems," *kan* 干.
6. For example, the character representing the midday hour of "Horse" is written as 午, while the character representing the animal is 馬.
7. *Gozen* 午前 and *gogo* 午後.
8. In Japanese, 明け六つ *akemutsu* and 暮れ六つ *kuremutsu*.
9. The *Classic of Changes* (*Yi Jing* 易經), popularly known as *The Book of Changes*, and in an older transliteration method as the *I-Ching*.
10. The anonymous author of *Jūniji shōko no kazu* and Terajima Ryōan in *Wakan Sansai Zue*, scroll 7, mention the origin of the multiplication; Robertson and Mody explain the sequence by a similar multiplication of the number nine, but they do not point to the *Classic of Changes* as a source and fail to explain why the multiplication ends at six and then starts over again. Brandes dismissed what he called "numerical mysticism" and sought to find a more "rational" explanation, all while admitting that he was completely ignorant in Japanese, Chinese, or Korean, and had only superficial knowledge of astronomy. "The Secret of Japanese Clock Dial," pp. 534–35.
11. 考証学. This school of thought originated in Qing China and became influential in Japan from the late seventeenth century on. Maruyama Masao dubbed this approach "historicist." Taking a more nuanced approach in *On Their Own Terms*, Benjamin Elman showed how such philological research laid the foundation for a more empirical approach for investigating nature. See also Ng Wai-ming, *The I Ching In Tokugawa Thought and Culture.*
12. In his discussion of the unit of the "day" astronomer and mathematician Nakane Genkei (1662–1733) made a clear distinction between the time of "astronomical day" and "human day." *Jikokuron*.
13. See, e.g., *Wakan sansai zue* and *Jūniji shōko no kazu*.
14. 天智天皇 (626–672, emperor 668–672).
15. 時香盤.
16. Golovnin, *Zapiski flota kapitana Golovnina*, p. 214.
17. 刻.
18. The earliest record of a man-made timepiece in Japanese history is found in an

eighth-century historical chronicle, the *Nihon shoki*, which states that in the sixth year of the reign of emperor Saimei (AD 660), "Hitsugi-no-Miko built a clepsydra for the first time, to let people know the time." Then, "on the first day of the fourth month of the tenth year of emperor Tenchi's reign [AD 671], the clepsydra was placed on a platform, and by setting drums and bells in motion, it struck the hours for the first time." Sakamoto Tarō et al., ed., *Nihonshoki*, vol. 4, p. 358, and vol. 5, p. 56, respectively.

19. 節気.
20. The length of the solar year does not divide neatly into 24. The accurate value of *sekki* that was used by Tokugawa-period calendar makers was 15.21 days. Minamoto, *Tenji meikai*, pp. 3–4.
21. Another problem was that the seasons did not fit the correspondence theory of *yin-yang* and the five phases. The four seasons seem to perfectly fit four out of five phases: the blossoming spring corresponded to wood, the hot summer to fire, the golden leaves of autumn to metal, and the rainy winter to water—yet there was no season to correspond to the earth phase. To solve this, a fifth, irregular "earth" season, called *doyō* (土用), was added, which occurred during the last eighteen days of each of the other four seasons. Obviously, this pseudo-season did not correspond to any climatic phenomena, and was rather a calendrical device designed to bridge inconsistencies inadvertently created by the cosmological system.
22. 候.
23. Kusa Dōin, *Shichijūnikō-shō*; *Wakan Sansai Zue*, fourth scroll, *Jikō rui*.
24. Sōma et al., "Units of Time in Ancient China and Japan."
25. Okada, *Kyūreki no yomihon*, p. 143.
26. Yamamoto, *Shichijūnikō idōkō*.
27. The fact that the moon's cycle is roughly twenty-nine and a half days was widely known since the antiquity. In the Tokugawa period it was accepted that lunar month is 29.53 days, close to the modern value of 29.53059 days. These values were recorded by Minamoto, *Tenji meikai*.
28. Jap. *urūzuki* 閏月.
29. *Wakan Sansai Zue*, fourth scroll, *Jikō rui*.
30. Nathan Sivin argues that, when referring to the yearly chart, we should translate the term 暦 as "almanac" rather than "calendar," to convey the divinational significance of some of the included events (*Granting the Seasons*, pp. 38–40). I disagree with this distinction, because (a) the purpose of the yearly schedule as a basis for social coordination is the same, (b) numerous Tokugawa calendars did not include information about auspicious events, while our own modern calendar is hardly devoid of religious elements, and (c) the English word "almanac" acquired its divinational connotation only in the twentieth century.
31. For more extensive information about the structure of Tokugawa calendars, see works by Gerhard Leinss.
32. 年号. The length of each era was not uniform, and could range from a few months to several decades. Only starting with the Meiji period were the eras aligned with reigns of different emperors. Before that, the beginning of a new era was not necessarily determined by imperial succession, but could follow the observation of astronomical anomalies, major earthquakes, or epidemics, or result from favorable astrological calculations. The names given to those eras often included characters of positive values like "virtue" and "correctness," often combined with terms like "eternity" or "abundance." Satō, "Comparative Ideas of Chronology," p. 286.

33. 干支.
34. *Rekichū* 暦註.
35. *Uranai* 占い.
36. For divinational and hemerological practices, see works by Matthias Hayek and Hayashi Makoto.
37. Shōtei, *Rekijitsu chūkai*.
38. *Rōkoku-hakase* 漏刻博士, *Tenmon-hakase* 天文博士, *Koyomi-hakase* 暦博士, and *Onmyō-hakase* 陰陽博士.
39. The Satsuma domain, at the far southwest of the archipelago, was allowed to keep its own hours given the dramatic difference in their length from those of the country's center. Okada, *Kyūrekino yomihon*, pp. 60–78.
40. *Ryaku-goyomi* 略暦.
41. *Daishō* 大小暦.
42. 橘南谿 (1753–1805). Tachibana Nankei, *Tōyūki*.
43. In "Literacy Revised," Peter Kornicki stresses that (il)literacy cannot be inferred from the lack of writing in certain documents.
44. Watanabe Toshio, "*Edojidaino rekihon hankōsū*."
45. Okada, *Kyūrekino yomihon*, p. 61.
46. *Toki no kane* 時の鐘. Urai, *Edo no jikoku to toki no kane*, pp. 36, 173.
47. Tsunoyama, *Tokei no shakaishi*, p. 71.
48. Golovnin, *Zapiski flota kapitana Golovnina*, p. 215.
49. Some sources claim that by the end of the seventeenth century there were about thirty thousand time-bells in Japan. See, for example, Hirai, *Tokei no hanashi*, p. 114.
50. Urai, *Edo no jikoku to toki no kane*, pp. 172–92; Hashimoto Manpei, *Nihon no jikoku seido*, pp. 157–70.
51. Based on Pierrre Charlvoix's *Histoire du Japon*. Bedini, "Oriental Concepts of the Measure of Time," p. 454.
52. 芭蕉 (1644–1694) and 曾良 (1649–1710).
53. Bashō, *Oku no hosomichi*.
54. Takahashi, *Edo no soshō*.
55. Tsunoyama, *Tokei no shakaishi*, p. 114.
56. Morishita, "Time in Early Modern Local Community," p. 72.
57. Ministry of Justice Research Dept., ed., *Tokugawa minji kanreishū*, pp. 113–15.
58. Morishita, "Time in Early Modern Local Community," p. 75.

CHAPTER TWO

1. St. Xavier (1506–1552). 大内義隆 (1507–1551). This clock has not survived to the present day, but both the Jesuit accounts and the records of Lord Ōuchi Yoshitaka mention such a gift. Yamaguchi, *Nihon no tokei*, p. 11; Tamura, *Ibaraki no tokei*, p. 121.
2. 織田信長 (1534–1582).
3. Cooper, *They Came to Japan*, p. 96.
4. As one Japanese source put it, European clocks "did not distinguish between long and short hours." Hirai, *Tokei no hanashi*, p. 129.
5. 徳川家康 (1543–1616) received the title of shogun in 1603. He abdicated in 1605 in favor of his son, but maintained power until his death. The ironsmith is 津田助左衛門. According to the *History of Owari Domain* (*Owari-shi* 尾張志).
6. It is important to note that the *History of Owari Domain* was written in 1843, more than two hundred years after the events described here, making it of questionable

reliability. Nevertheless, even if Tsuda was not the first clock-maker, this story is representative of the emergence of the profession in the late sixteenth century.

7. 和時計.
8. The latter was used in Chinese texts such as *Zhou bi suan jing* and *Huainan zi*. Hirai, *Tokei no hanashi*, p. 21.
9. 自鳴鐘.
10. Hashimoto, *Nihon no Jikoku Seido*, p. 84.
11. Satō, *Rekigakushi taizen*, p. 200.
12. *Yukiwa* 雪輪. This was a gear regulating the number of strikes of the alarm.
13. *Gyōjiwa* 行司輪.
14. *Tenpu* 天府.
15. *Ken* 剣, and *koma* 駒.
16. Makieshi, *Jinrin kinmōzui*, "Artisans."
17. Ihara Saikaku, *Nippon Eitaigura* 日本永代蔵 (1688) and *Kōshoku ichidai otoko* 好色一代男 (1684).
18. Takeda Izumo et al., *Kanadehon chūshingura* 仮名手本忠臣蔵 (1747).
19. 中根元圭 (1662–1733).
20. Nakane, *Jikokuron*.
21. Ibid.
22. 掛け時計.
23. 櫓時計. Mody and Robertson call these "lantern clocks," referring to the Dutch timepieces known by that term.
24. 枕時計. Mody and Robertson call these "bracket clocks," referring to the European prototype.
25. Hirai cites *Tokugawa Jikki* (Records of Tokugawa family), which states that the third shogun, Iemitsu (ruled 1623–1651), owned many of those. Hirai, *Tokei no Hanashi*, p. 124.
26. Nishikawa Joken, *Nagasaki Yawasō*, p. 80; Kyūshidō Shūjin, *Shōchū jishingijimō*; Kondō Katsuno shows evidence that some pillow clocks were produced by Japanese clock-makers as well. Kondō Katsuno, *Saishoki makuradokei ni tsuite*, p. 1.
27. *Manjū dokei* 饅頭時計.
28. 袖時計.
29. 印籠時計.
30. 割駒.
31. Hirai, *Tokei no hanashi*, p. 26.
32. 尺時計.
33. Of course, *shaku* clocks had to be wound every day, since the mechanism could only go down, with no automatic feature to return the weight to the top. But since daily winding the clock was a common feature of most late seventeenth- and early eighteenth-century clocks, this was not seen as a major inconvenience.
34. Transcribed in Kondō, *Furiko en gurafu shiki mojiban kakedokei*, p. 5.

CHAPTER THREE

1. Nakayama, *Nihon no Tenmongaku*, p. 45.
2. 宣明暦.
3. 保科正之 (1611–1673).
4. 徳川家光 (1604–1651).
5. 徳川家綱 (1641–1680).
6. 渋川春海 (1639–1715).

7. 授時暦. This title was not just poetic superfluity, but rather expressed an underlying ideological perspective according to which a benevolent government is responsible for supplying citizens with a proper calendar in order to enable productive agricultural activity. See Sivin, *Granting the Seasons*.
8. 容螺山 Park An-ki Nasan. According to Shunkai's student Tani Shinzan 谷秦山 (1663–1718) in *Shunkai Sensei Jikki*. Cited in Nakayama, *Nihon no Tenmongaku*, p. 48.
9. Christian convert Hayashi Kichizaemon (林吉左衛門) practiced Western astronomy until he was charged with propagating Christianity and executed in 1645. A student of his, Kobayashi Yoshinobu (小林義信 1601–1684), escaped that fate, being imprisoned for "only" 21 years, which he spent writing his *Nigi ryaku setsu* (Abbreviated explanation of the two spheres 二儀略説), based on *De Sphaera*, by Pedro Gomez. Nakayama, *A History of Japanese Astronomy*, p. 99.
10. After the tortures he endured for trying to infiltrate the Christian faith into Japan, Chiristóvão Ferreira (c.1580–1650) abandoned his religion and decided to stay in Japan, marrying a Japanese woman and taking a Japanese name, Sawano Chūan (沢野忠庵). Although fluent in spoken Japanese, he could not express his ideas in writing and dictated his (somewhat general) knowledge of Ptolemaic astronomy to a Japanese scholar, Mukai Genshō (向井元升), who rendered it into classical Chinese, naming it *Kenkon Bensetsu* 乾坤弁説 (Explanation of heaven and earth). Mukai Genshō also contributed an introduction and commentaries to the text, in which he criticized what he perceived to be the shortcomings of Western astronomy. Nakayama, *A History of Japanese Astronomy*, 88–98.
11. 天経或問 (*Questions about Heavens*, Jap. *Tenkei wakumon*) was written by You Yun (游芸), a student of Jesuit missionary Emanuel Diaz but not a professional astronomer himself. Due to the illegal nature of its arrival to Japan, it is not clear when exactly it was brought and when the first copies of it were printed. It is clear, however, that by the last decade of the seventeenth century it was widely read, to the point that it was extensively cited in the popular encyclopedia *Wakan san sai zue*, first published in 1713.
12. In *Books and Boats*, Ōba Osamu describes the vast scope of officially forbidden literature that nevertheless was smuggled into Japan and circulated in bookstores.
13. See, for example, Shibukawa Shunkai, *Tenmon keitō*, pp. 118, 125.
14. According to Shunkai's student Tani Shinzan, *Shinzanshū*. Cited in Nakayama, *Nihon no Tenmongaku*, p. 75.
15. From a geocentric perspective, the tropical, or solar, year, is the time it takes the sun to return to the same position relative to the earth. The tropical year is often defined by successive solstices or equinoxes. The sidereal year, on the other hand, is the time it takes the sun to return to the same position relative to the fixed stars.
16. Sivin, *Granting the Seasons*, pp. 297–304.
17. This is crucial to the decision how to place months in a year—whether to simply alternate months of 29 and 30 days, or determine their position according to the actual conjunctions relative to the sun. *Granting the Seasons*, p. 368.
18. What Sivin calls "proto-trigonometry." *Granting the Seasons*, pp. 66–67.
19. Although many of them were completed only after the reform.
20. The legendary water clock designed by Su-Song was not among the instruments.
21. For detailed explanation of the instruments, see Sivin, *Granting the Seasons*, pp. 179–211.
22. *Yang yi* 仰義.
23. *Jien yi* 簡義.

24. *Bai ke huan* 百刻環.
25. He studied mathematics under the famous Japanese mathematician Seki Takakazu 関孝和 (1642–1708), and medicine under Okanoi Gentei (岡野井玄貞), through whom Shunkai met the abovementioned Korean scholar Razan. Yamazaki Ansai 山崎 闇斎 (1619–1682) began his career as a Buddhist monk and gradually changed his philosophy to focus on a combination of neo-Confucianism and Shinto. For more on Yamazaki Ansai, see Ooms, *Tokugawa Ideology*, and Tucker, "Religious Dimensions of Confucianism: Cosmology and Cultivation."
26. Hayashi, *"Igo to tenmon."*
27. 大和暦. Compare to Ansai's own *Yamato Learning* (大和小学), and a later adaptation of Chinese material medica by Japanese scholar Kaibara Ekiken (貝原益軒) titled *Yamato Honzō*. On the latter, see Federico Marcon's *The Knowledge of Nature and Nature of Knowledge.*
28. Okada, *Kyūreki no yomihon*, pp. 142–65.
29. 里差 Chi. *Licha* Jap. *Risa.* The term literally means "distance in [length units of] *li/ri*" (里). 里 could be roughly translated as "miles" or "leagues," but the precise length of this unit varied over time and differed between China, Japan, and Korea. In modern-day dictionaries in China, one *li* equals about 500 meters; in Korea, it traditionally equates to about 400 meters, but in Japan, it is almost four kilometers. These definitions, however, rely on nineteenth-century calculations, and do not reflect the conceptualization of the unit in the seventeenth, let alone the thirteenth, century.
30. In certain cases it could indicate parallax, Jap. *shisa* 視差.
31. According to Nishimura Tōsato in *Sokuryō sho ki zu*. Also cited in Watanabe Toshio, *Kinsei Nihon Tenmongakushi: Kansatsu Gijutsu*, p. 556.
32. Literally "law of contractions and expansions" 消長法 Jap. *Shōchōhō*; Chi. *Xiaozhangfa*.
33. Shunkai took one year to be 365.25 days.
34. *Byō* 秒 and *bu* 分, which simply mean "minuscule" and "part." These units later evolved into what we know today as seconds and minutes. Watanabe, *Kinsei Nihon Tenmongakushi*, pp. 73–74.
35. The system was called the Jōkyō calendar 貞享暦.
36. As previously noted, the Satsuma domain, on the southern island of Kyushu, was allowed to keep different time. Although this was justified by the fact that the domain lay significantly westward of Kyoto, it also reflected the era's politics. The Shimazu clan who ruled Satsuma were in opposition to the Tokugawas when the latter assumed the shogunate in the early seventeenth century. The clan was one of the richest among the domains, second only to Kaga in its income. It also controlled the Ryūkyū islands (present-day Okinawa) and most of the foreign trade was conducted on its territory.
37. 関孝和 (1642–1708), 建部賢弘 (1664–1739), 中根元圭 (1663–1733). See Horiuchi, *Japanese Mathematics in the Edo Period*.
38. In his examination of *Granting the Seasons*, Sivin relied heavily on the explanations written by Takebe Katahiro, whose treatise on its mathematical methods remains the most detailed explication of the system.
39. 西洋新法暦書. Horiuchi, *Japanese Mathematics in the Edo Period*, p. 224.
40. *Discussions on the Principles of Astronomy* (天文義論 *Tenmon Giron*), in 1712, and *Collected Explanations about the Two* [terrestrial and celestial] *Spheres* (両儀集説 *Ryōgi shūsetsu*), in 1714.

41. Seeing that the ban was lifted before any recorded engagement of Japanese mathematicians with imported books, Horiuchi insists that it was Yoshimune's own decision to lift the ban. Nevertheless, his close relationship with Takebe Katahiro suggest that the latter exercised not a small amount of influence on the shogun. *Japanese Mathematics in the Edo Period*, pp. 221–31.
42. *Lisuan quanshu* 暦算全書; Jap. *Rekisan zensho*. Mei Wending 梅文鼎 (1633–1721).
43. *Chongzhen Lishu* 崇禎暦書; Jap. *Suitei rekisho.*
44. *Lixiang kaocheng* 暦象考成; Jap. *Rekishō kōsei.*
45. *Lingtai yixiang zhi* 靈台儀象志; Jap. *Reidai Gishōshi.*
46. For more information on the translation of Western sciences in sixteenth- and seventeenth-century China, see Engelfriet, *Euclid in China*, Zhang, *Making the New World Their Own*, and Elman, *On Their Own Terms*.
47. Watanabe Toshio, *Kinsei Nihon Tenmongakushi*, p. 91.
48. 享保. The reforms were named after the era during which they were initiated.
49. Such complaints appear in Ōshima Shiran, *Tokei Kō*.
50. *Tokugawa Jikki*, p. 293; also cited in Nakamura, *Edo no Tenmongakusha Hoshizora o Kakeru*, p. 64.
51. 西川正休 (1693–1756).
52. Abe Yasukuni 安部泰邦 (1711–1784). Tsuchimikado 土御門.
53. *Jujikan* 授時简.
54. Abe Yasukuni, *Rekihō Shinsho*, pp. 160–61, 180.
55. *Senshigi* 暹胝儀.
56. *Rekihō Shinsho*, p. 160
57. *Rekihō Shinsho*, p. 161.
58. 修正. Hisashi Uehara et al., ed., *Tenmon rekigaku shoka shokanshū*, pp. 19–23.
59. Nishimura Tōsato, *Honchō Tenmonshi*, cited in Watanabe Toshio, *Kinsei Nihon Tenmongakushi*, pp. 144–45.
60. *Honchō Tenmonshi*, pp. 144–45.
61. 麻田剛立 (1734–1799).
62. Gōryū's early interest was in anatomy. His medical experiments, as well as his astronomical observations, are well recorded in his letters to the famous scholar Miura Baien (三浦 梅園 1723–1789). See, for example, Hisashi Uehara et al., ed., *Tenmon rekigaku shoka shokanshū*, pp. 44–45, Letter to Miura Baien from 16th day, 10th month, unknown year.
63. As a salaried samurai, leaving the position without permission was akin to deserting, and thus subject to prosecution. Gōryū's former employer, who was also a childhood friend, did not press charges, but also was not willing to pay salary for an absent employee, dooming Gōryū to years of poverty.
64. Benjamin Elman, *On Their Own Terms.* The process of translating Western sciences into classical Chinese is in itself revealing of the complexity of cultural assumptions embedded in scientific principles. Jesuits were translating not only language, but also culture. Qiong Zhang has written extensively on translations between different worldviews—geographical and metaphysical. She shows not only the struggle to find appropriate terms in Chinese to match Latin terms, but also the opposite—how Jesuits attempted to understand the concept of *qi*. See *Making the New World Their Own* and "Demystifying Qi." Similarly, in *Euclid in China* Peter Engelfriet shows how apparently straightforward geometrical axioms were nevertheless laden with cultural meaning.

65. Uehara et al., ed., *Tenmon rekigaku shoka shokanshū*, p. 31. Letter to Miura Baien from 22nd day of the 7th month, year unknown.
66. *Reidai gishōshi*, scroll 4.
67. Cited in Watanabe, *Kansatsu Gijutsu*, p. 560.
68. Matsudaira Sadanobu (松平 定信 1759–1829).
69. "The elders" (老中 *rōju*) were those holding the highest administrative positions. Although there were several "elders," Sadanobu quickly became the most powerful.
70. The Kansei (寛政) era lasted from 1789 to 1801.
71. 間重富 (1756–1816), 高橋至時 (1764–1804).
72. *Lixiang kaocheng houbian* 暦象考成後編; Jap. *Rekishō kōsei gohen*, written by Ignatius Koegler. Japanese astronomers often referred to this book simply as *The Sequel* 後編.
73. Hazama Shigeyori 間重新, *Senkō taigyō sensei jiseki ryakki*, p. 455.
74. Watanabe Makoto, *Endō Takanori wo chūshin ni okonawareta Kagahan no gijutsubunka no kenkyū*, p. 51.
75. 時中法. Sometimes spelled as 侍中法.
76. Nakayama, *Nihon Tenmongaku*, p. 120.
77. Sources disagree concerning the identity of the inventor of the pendulum clock, and at least one source from the 1830s attributes it to Asada Gōryū. Watanabe Toshio, however, convincingly argues that it was Hazama Shigetomi who invented the mechanical device, while Asada Gōryū used only a pendulum. Watanabe, *Kansatsu Gijutsu*, pp. 559–61.
78. Suiyōkyūgi 垂揺球儀.
79. Hazama Shigeyori, *Senkō taigyō sensei jiseki ryakki*, pp. 452–53.
80. Nishimura Tachū, *Shitendai sho ki zu*.
81. 象限儀 and 子午線儀. The name *shigosengi* literally means the "Rat-Horse-line device," referring to the animal signs that represented not only the hours but also the directions, in this case the north and the south.
82. Shibukawa Kagesuke, *Kansei Rekisho*, pp. 425–29.
83. 象限子午線垂揺球儀.
84. Watanabe Toshio argued that Asada Gōryū "discovered" Kepler's third law before it was formally introduced to Japan. Gōryū, however, did not come up with the same formula, even though he verbally described that such a relationship exists. *Kinsei Nihon Tenmongakushi*, p. 179.
85. The Japanese expression is天理自然の数. *Suikyūseigi*.
86. In his *Knowledge of Nature and Nature of Knowledge*, Marcon makes a similar claim, that the modern notion of "nature" developed in Tokugawa Japan in relation to sets of scholarly practices—in his case, in the field of plant and animal studies.
87. *Suikyūseigi*.
88. Ibid.
89. Ibid.
90. Ibid.
91. Ibid.
92. In *Restless Clock*, Jessica Riskin showed that although the metaphor of the clock guided centuries of European scientific thought, the particular meanings people ascribed to this metaphor changed with the transformation of intellectual trends.

CHAPTER FOUR

1. Yoshitoki detailed his troubles with the eclipse calculations in a letter to fellow astronomer Hazama Shigetomi in 1798. Cited in full in Uehara, *Takahashi Kageyasu no kenkyū*, p. 144.
2. 里.
3. 尺.
4. This opinion is explicitly expressed by Baba Sajūrō (馬場佐十郎 1787–1822), a young translator who was struggling to translate foreign units into comprehensible terms. See *Doryōkō*, 1810, cited in Uehara, *Takahashi Kageyasu no kenkyū*, p. 160.
5. Ōtani, *Tadataka Inō*, p. 46.
6. Japanese astronomers had concerns about the disarray of standards similar to those of European astronomical communities around the same time. See Ken Alder, *The Measure of All Things*.
7. Baba Sajūrō, *Doryōkō*, 1810, cited in Uehara, *Takahashi Kageyasu no kenkyū*, pp. 160–61. Although this source was written some ten years later, it is still representative of the confusion about the units of length at the turn of the century.
8. See, for example, Takahashi Yoshitoki, *Seiyōjin Rarande rekisho kanken*, pp. 168–69.
9. Takahashi Yoshitoki, *Seiyōjin Rarande rekisho kanken*, p. 169.
10. Egbert Buys, *Woordenboek*, "Aardkloot," pp. 31–39.
11. We know for sure that Yoshitoki saw this table because he referred to it in his later writings and even copied it in its entirety. *Seiyōjin Rarande rekisho kanken*, p. 169.
12. Today we do acknowledge the existence of a little bulge around the equator, albeit nothing as large as suggested by Buys's book. In late eighteenth-century Europe, however, the existence and the dimensions of this bulge were treated as theoretical possibilities deriving from Newton's theory of gravitation, not as confirmed fact. For the discussion of these debates, see Terrall, *The Man Who Flattened the Earth*, esp. p. 57.
13. Takahashi Yoshitoki, *Seiyōjin Rarande rekisho kanken*, p. 170.
14. 伊能忠敬 (1745–1818).
15. Murai Masahiro, *Ryōchi shinan* (1732); Mao Tokiharu, *Kikubuntōshū* (1722); and Shimada Dōkan, *Chōkenbengi* (1725), all of which were listed in Inō's library.
16. Inō Tadataka described his own experiences in a treatise that vehemently criticized one particular attempt to establish an astronomical system consistent with Buddhist cosmology. *Bukkoku Rekishohen Sekimō*, cited in Uehara, *Takahashi Kageyasu no kenkyū*, p. 146.
17. In terms of surveying and observational methods, this mission is very similar to Maupertuis's expedition to Lapland, described by Mary Terrall in *The Man Who Flattened the Earth*. As for the goals of the expedition, however, this case is reminiscent of Delambre's and Méchain's measurement of France's portion of the meridian, described by Ken Alder in *The Measure of All Things*.
18. Inō Tadataka, *Enkai chizu hanrei, narabini, chizu fuki tōto riteiki*, cited in Uehara, *Takahashi Kageyasu no kenkyū*, p. 148.
19. 大野弥三郎 (規行).
20. 戸田藤三郎 (忠行).
21. Uehara, *Takahashi Kageyasu no kenkyū*, pp. 160–61.
22. In this sense, Inō's expedition was nothing like "lightly equipped, underfinanced, more or less solitary" expedition of Robert Herman Schomburgk, described by Graham Burnett in *Masters of All They Surveyed*. Citation from p. 3.
23. Ōtani, *Tadataka Inō*, pp. 79–81.

24. Compare to the current value: 110.98 km.
25. Cited in Uehara, *Takahashi Kageyasu no kenkyū*, p. 149.
26. In the original text, he is referred to by his nickname, "Master Higashioka."
27. Takahashi Kageyasu, *Ryō'ō den*, cited in Uehara, *Takahashi Kageyasu no kenkyū*, p. 150.
28. I have explored the process of Yoshitoki's nonverbal interpretation of Lalande in the article "Before Words: Reading Western Astronomy in Early 19th Century Japan."
29. Takahashi Kageyasu, *Ryō'ō den*, cited in Uehara, *Takahashi Kageyasu no kenkyū*, p. 160–61.
30. See David Howell, *Geographies of Identity in Nineteenth-Century Japan*. There are also many Edo-period sources depicting the "barbaric" peoples of the northern island. See, e.g., *Illustrated Explanations of Northern Ezo*, by Mamiya Rinzō and Murakami Teisuke.
31. Inō Tadataka, *Sokuryō nikki*.
32. A good example is *Ryōchiden shuroku*, written by Inō's student Watanabe Shin.
33. Suiyōkyūgi 垂揺球儀. See chapter 3.
34. Inō Tadataka, *Enkai chizu hanrei, narabini, chizu fuki tōto riteiki*, cited in Uehara, *Takahashi Kageyasu no kenkyū*, p. 148.
35. Yoshitoki's letter to an interested amateur in 1800. Hisashi Uehara et al., ed., *Tenmon rekigaku shoka shokanshū*, pp. 68–74.
36. See, for example, Inō's *Nihon keiido jissoku*.
37. 麻田立達 (1771–1827). Eldest son of Shigetomi's and Yoshitoki's teacher, Asada Gōryū.
38. Letters from Kansei 10 (1798), cited in full in Watanabe, *Kinsei Nihon Tenmongakushi*, p. 607.
39. Letters from Kansei 10 (1798), cited in full in Watanabe, *Kinsei Nihon Tenmongakushi*, p. 609.
40. Which indeed happen to Hazama Shigetomi, who traveled all the way to Kyushu to observe a total lunar eclipse, but failed to do so because of the weather.
41. Watanabe, *Kansatsu Gijutsu*, p. 611.
42. Latour, in "Drawing Things Together," attributes primacy to "inscriptions" of actual maps in constituting social claims. Burnett in *Masters of All They Surveyed* also focused on the primacy of the map. My point here is to show that there is a long conceptual process *before* the actual inscription. This process is not necessarily singular, confined to one specific explorer, but rather rooted in a set of practices and communal conceptual development, even though the "community" might be very small at any given point.
43. 間宮林蔵 (1775–1844).
44. Similarly to Bruno Latour's La Pérouse, Mamiya Rinzō's intentions were, if not imperialistic, then at least done in the interest of the government. Brett Walker gives an outstanding description of Mamiya's travels and the political impact of the surveying of the "barbarian" north. Taking a Latourian approach, Walker attributes this achievement almost solely to Mamiya, claiming that he brought about a firmer control of the lands by "placing them on the grid of longitudes and latitudes." Yet Mamiya never made astronomical observations, and the results of his survey could be placed on the grid of longitudes and latitudes only after they were incorporated into Inō's maps, which did *not* originate in imperialistic motivations. "Mamiya Rinzō and the Japanese Exploration of Sakhalin Island."
45. Golovnin, *Zapiski flota kapitana Golovnina*, p. 173.

46. In this method, longitude could be found by measuring the distance between the moon and a preselected celestial body, and then comparing it to a table that provided the distance between these two bodies as observable from a location of known longitude.
47. Takahashi Kageyasu 高橋景保 (1785–1829).
48. Edo, thus, was what Latour defines as a *center of calculation*—a nexus of activity to which collected information and materials flow back, and around which social networks of scientists revolve. Latour, *Science in Action*, especially chapter 6, "Centres of Calculation."
49. As suggested by Latour in *Science in Action*.
50. Dai Nippon 大日本. Yonemoto in *Mapping Early Modern Japan* makes an excellent case for how travels, actual and imaginary, contributed to the creation of this notion of the unified entity of Japan. Here I want to stress the material (the clock) and the conceptual (space as time) aspects that contributed to the emergence of the notion of "Japan" as a unified entity.
51. The exact number of domains, or *han* 藩, fluctuated throughout the Edo period, but overall there were around 250 separate domains of various sizes, which varied greatly in their economic and cultural conditions.
52. Nakagawa Goroji, for example, was actually abducted by a Russian ship and spent several years on the continent. Upon his return to Japan, he helped disseminate a new vaccination method he learned in a Russian village.
53. *Rōshia shinto peteruburugu no zu*.
54. *Shintei Bankoku zenzu* and *Shinsen sōkai zenzu*.
55. At the same time, he undertook an extensive study of the Manchu language, as well as of Russian customs.
56. Watanabe, *Kansatsu Gijutsu*, p. 406. Hisashi Uehara maintains that Mamiya Rinzō meant no harm to Kageyasu and that he submitted his report due to his own fear of the possible repercussions of interaction with a foreigner. Uehara nonetheless agrees that Mamiya's report probably caused the secret police to open a file on Kageyasu and investigate his connections with Siebold even before the actual discovery of the maps. Uehara, *Takahashi Kageyasu no kenkyū*, pp. 314–18.
57. Uehara, *Takahashi Kageyasu no kenkyū*, pp. 336–37.
58. Ibid., pp. 337–38.
59. Ibid., p. 340.
60. The map actually says "Takahashi Sukezaemon," which was Kageyasu's real name. "Kageyasu" was a pseudonym, which means "Shadow Keeper," referring to the shadow of the gnomon that historically was considered to be symbolic for astronomy. In the same manner, his brother Kagesuke's pseudonym means "Shadow's Assistant"; his father Yoshitoki's name means "Reaching Time"; Inō Tadataka's pseudonym (rarely used in the literature today, but almost exclusively used by astronomers of his time) is "Kageyū"—"Relying on the Shadow," etc.
61. As La Pérouse claimed. Morris-Suzuki, "Indigenous Knowledge in the Mapping of the Northern Frontier Regions," p. 138.
62. In the chapter on navigation, I talk about his misunderstanding of the Japanese scholarly scene.
63. In a letter to Siebold. Cited in full in original German in Otani, *Tadataka Inō*, p. 150.
64. The political importance of the temporal center that defined the geographical continuum is well seen in the mid-nineteenth-century disputes concerning the placement of the prime meridian. See Peter Galison, *Einstein's Clocks, Poincaré's Maps*.

65. Suzuki Junko claims that the cartographical project described in this chapter took place because "the time was ripe." "Seeking Accuracy," p. 132. However one might wish to interpret that, it was specific astronomical motivations that set this project in motion.

CHAPTER FIVE

1. Literally "law of contractions and expansions" 消長法 Jap. *Shōchōhō*. Today we accept the fact that the length of the solar year is constantly changing, yet the rate accepted today is significantly lower than the one established by Gōryū.
2. Watanabe, *Kinsei Nihon Tenmongaku*, 202–14.
3. Today the character 分 is pronounced *pun/bun*.
4. See for example a letter from Takahashi Yoshitoki to Hazama Shigetomi, cited in Uehara, *Takahashi Kageyasu no Kenkyu*, pp. 144–45.
5. See chapter 4.
6. The invention of the marine chronometer in the West is often described as being motivated by a very real and practical problem—the loss of ships and their cargo due to the inability to correctly calculate their positions. See William Andrews, ed., *The Quest for Longitude*.
7. 本多利明 (1743–1821).
8. Mt. Asama sits northwest of Tokyo, on the eastern edge of the area known today as the Japanese Alps. It is still an active volcano that erupts regularly.
9. Especially noteworthy is *A Secret Plan of State Craft* (*Keisei hisaku* 経世秘策), written in the Kansei period (between 1789 and 1800). We should note, however, that this "secret" plan was widely published and discussed. For extended discussion of his utopian vision see Annick Horiuchi, "Honda Toshiaki (1743–1820) *ou l'Occident comme utopie*," in *Repenser l'ordre, repenser l'héritage: Paysage intellectuel du Japon (XVIIe -XIXe siècles)*, ed. F. Girard, A. Horiuchi, and M. Macé (Droz, 2002), pp. 411–48.
10. Recorded by Katsuragawa Hoshū (1751–1809) as *Hokusa Bunryaku*.
11. As in his claim that "the ruler of Rome, the capital of Italy, is the emperor of the whole of Europe." *Seiiki monogatari*, p. 98. The source of this mistake is not clear. One possibility is that Toshiaki was confused by the term "Holy Roman Empire." Another source of this misinterpretation might be his reliance on Jesuit sources from China, which represented the Pope as the strongest person in Europe.
12. *Seiiki monogatari*.
13. Horiuchi treats several of Toshiaki's writings as utopian. "Honda Toshiaki (1743–1820) *ou l'Occident comme utopie*."
14. *The Tales of the West* mentions the trio "astronomy, geography, and navigation" on almost every page. Honda Toshiaki uses the word "study" (*gaku* 学) or "way" (*dō* 道) to signify particular scholarly disciplines. However, like the many other terms that undergo a transformation in meaning in Toshiaki's writings, the terms are used in a way that resembles the modern word "science." *Seiiki monogatari*.
15. *Seiiki monogatari*, 98
16. Horiuchi described this as "geographical determinism" (*déterminisme géographique*). Horiuchi, "Honda Toshiaki (1743–1820) *ou l'Occident comme utopie*," p. 425.
17. See Montesquieu's *Persian Letters* and *The Spirit of the Laws*.
18. *Seiiki monogatari*.
19. Yoshitoki's letters in *Tenmonrekigaku shoka shokanshū*.
20. Shibukawa Kagesuke, *Zoku kaichū funamichi kō*.

21. Honda Toshiaki, *Tokai shinpō*.
22. Iida Yoshirō, "*Tokai shinpō ni okeru kōkaigaku*," pp. 1–25.
23. He distinguished between coastal sailing (*chijō* 地乗) and open sea sailing (*chūjō* 沖乗).
24. *Tokai shinpō*, pp. 263–65.
25. Unfortunately, Honda Toshiaki did not follow Inō Tadataka's method closely enough. Whereas the latter was observing the visible positions of the satellites of Jupiter, the timings of which could be predicted and recorded in a table, Toshiaki choose instead to focus on a random star. This method could not have produced the desired longitude, since the time he measured was an apparent time (defined by the interval between sun's crossing of the local meridian and its return to the meridian on the next day), while the moment when a star was at its southernmost position was in sidereal time (the time it takes a particular star to return to its position relative to the earth—a sidereal day is roughly 23 hours and 56 seconds in our terms).
26. Ibid., 371.
27. Ibid., 373.
28. *Seiiki monogatari*, 139
29. Yoshitoki in a letter to Anazawa Utarō from 1801, in *Tenmonrekigaku shoka shokanshū*, p. 71.
30. The 360 degrees of the earth's circumference divided by 24 equal hours. Honda Toshiaki, *Tokai shinpō*, 235.
31. 時辰儀.
32. Aka Ivan Fyodorovich Kruzenshtern (1770–1846).
33. Edo.
34. Kruzenshtern, *Voyage around the world*, pp. 279–80. It is unclear what Kruzenshtern meant by the term "*Issis*." Astronomers were referred to as "*tenmongata*" and conducted their work at the Asakusa observatory, not in a temple. There were other officials—the "Tsuchimikado"—who interpreted astronomical data in hemerological terms.
35. Siebold, *Nippon*, p. 53.
36. On February 16 and 24, March 5 and 7, and April 14. Siebold, *Nippon*, pp. 79, 112, 133, 139, 187.
37. See chapter 4 for more on the "Siebold affair."
38. 渋川景佑 (1787–1856).
39. Shibukawa Kagesuke, *Reigen kōbo*, p. 564.
40. Cited in Watanabe, *Kinsei Nihon Tenmongaku*, p. 564.
41. Shibukawa Kagesuke, *Shōchū jishingi jimō*, p. 567.
42. Satō Issai's diary from 1841 cited in Nakamura Tsukō, "*Satō Issai no tokei kenkyū to bakumatsu tenmongata to no kōryū*," pp. 23–24.
43. Mahoney, "Christian Huygens: The Measurement of Time and of Longitude at Sea."
44. Takahashi Yoshitoki, *Angeria reki kō*.
45. Ibid.
46. Shibukawa Kagesuke, *Zoku kaichū funamichi kō*.
47. *Jishingi montō*. Although the piece is not dated, Nakamura Tsukō used the data mentioned in the text in order to calculate the year in question, and arrived at 1846. "*Shibukawa Kagesuke cho 'jishingimontō' no seiritsu nen no suitei*."
48. *Jishingi montō*.
49. Shubukawa Kagesuke uses "universal time difference" as a term for the equation

of time. Given that, traditionally, Japan celebrated New Years in early February, the seventh day of the first month is in the vicinity of February 11, which coincides with one of the peaks in the fluctuation of apparent time. Kagesuke probably intentionally chose a date that featured the largest possible difference between the apparent and mean solar day.

50. *Jishingi montō.*
51. 用時 *yōji* and 平時 *heiji*. The literal translation of the Japanese term *yōji* is actually "the time that is used," which assumes that this kind of time was in *common* use.
52. McCrossen, *Marking Modern Times.*
53. Iida Yoshirō, "*Kōkai yōhō setsuyaku nitsuite shōkai,*" p. 112.
54. Fujino Masahiro, *Kōkai yōhō setsuyaku.*
55. *Kōkai hitsudoku.*
56. Ibid.
57. Ibid.
58. *Kōkai yōhō setsuyaku.*
59. Ibid.
60. Vanessa Ogle describes this ideological struggle in *The Global Transformation of Time.*
61. The Ottoman Empire is often cited as an example of the forceful spread of Western style timekeeping and the idea of mean time. On Barak, *On Time*; Vanessa Ogle, *The Global Transformation of Time.*

CHAPTER SIX

1. 遠藤高璟 (1784–1864).
2. Motoyasu, "*Kagahan to 19 seiki no 'gijutsu bunka',*" p. 2.
3. 西村太沖 (1767–1835).
4. Nishimura Tōsato, *Sokuryō sho ki no zu.* See figures 1.3 and 3.2.
5. His official titles were *Sakuji Bugyō* (作事奉行) and *Fushin Bugyō* (普請奉行).
6. Time Known Through Sun's Shadows *Kichiji* 晷知時, and Sun's Shadows Timepiece *Kijiki* 晷時規. Watanabe Makoto, *Endō Takanori wo chūshin ni okonawareta Kagahan no gijutsubunka no kenkyū*, p. 59.
7. *Kanazawa jishō ki* cited in *Endō Takanori wo chūshin ni okonawareta Kagahan no gijutsubunka no kenkyū.*
8. Jap. *amari* 余.
9. 石黒信由 (1760–1836).
10. 本多利明 (1743–1821).
11. 究理学.
12. *Sūri* 数理, *suiho* 推步, *sokuryō* 測量. *Seiiki Monogatari*, 90.
13. Honda Toshiaki, *Seiiki Monogatari*, p. 118.
14. 三角風蔵 (1784–1868).
15. On Toshiaki's geographic determinism see chapter 5.
16. *Sokkibai* 測晷牌.
17. 前田斉広 (1782–1824).
18. *Sokkiban* 測晷盤.
19. Endō Takanori, *Takezawa goten soku koku go kibutsu yōhō.*
20. *Seijikoku* 正時刻.
21. Watanabe Makoto, "*Kanazawajō jōshō no yakuwari to jikoku seido no hensen,*" p. 64.
22. *Shōjiban* 正時版.
23. Endō Takanori, *Takezawa goten soku koku go kibutsu yōhō.*
24. Matsuzaki Toshio, *Edo Jidai no Sokuryōjustu*, 189.

25. Watanabe Makoto, *"Kagahan ni okonawareta henkaku no sokutei to sono mokuteki,"* p. 95.
26. Ibid., pp. 95–98.
27. Endō Takanori, *Hōkei Ninjū* (Directions are light, humans are heavy 方軽人重) cited in Onabe Tomoko, *Zettai Tōmei no Tankyū*, p. 65.
28. Although, in retrospect, the precise values calculated in these four instances, especially the first one, were a bit imprecise.
29. 黒川良案 (1817–1890).
30. Motoyasu Hiroshi, *"Kagahan to 19 seiki no 'gijutsu bunka',"* p. 3.
31. *Shikeigi* 眡景儀. Note that the first character no longer exists in Japanese.
32. *Yotsu no shirabe* よつのしらべ, or 四能導. Note that even though the latter variation is written in characters, it is not in classical Chinese but rather in Japanese-style writing that often uses characters for phonetic purposes.
33. 写法新術. The Union Catalog of Early Japanese Books lists the date of the manuscript as 1841 (Tenpō 12); however, Tomoko Onabe claims that the manuscript was only completed in 1850 (Onabe, *Zettai Tōmei no Tankyū*, p. 3). I examined the copy preserved at the National Academy of Science (Gakushiin), which consists of several different versions.
34. Onabe's *Zettai Tōmei no Tankyū* is the only book-length study of the treatise and offers the most thorough research on Takanori's views on depiction.
35. Endō Takanori, *Shahō shinjutsu*, Preface.
36. 真.
37. *Arinomama*, ありのまま.
38. Cited in Onabe, *Zettai Tōmei no Tankyū*, p. 100.
39. Method of assessment of the mind *shinsekihō* 心積法, method of assessment of things *bussekihō* 物積法, and method of assessment of observation *kansekihō* 観積法.
40. *Shahō shinjutsu*, Preface.
41. Ibid., vol. 3.
42. Ibid., vol. 3.
43. This may refer to the Sand Point Island off Alaska, or the Sand Island in Hawaii. Since Takanori included Hawaii on his map, I tend to think it was the latter.
44. Timon Screech provides an entirely different interpretation of this passage, according to which Takanori (whom Screech described as "a samurai from Kaga") seemed to be amazed at the sight of the clock, comparing the index hands to swords. Screech's reading, however, ignores the fact that the word "sword" in Japanese (*ken* 剣), was conventionally used to indicate an index hand. "Clock Metaphors in Edo Period."
45. *Kenjō onjiki yurai narabini yōhō no oboe.*

CHAPTER SEVEN

1. The relationship between the art of automata and conceptualization of nature is not at all unique to Europe. See, for example, Jessica Riskin's *Restless Clock*.
2. Saikaku's verse uses an ambiguous term, *morokoshibito*, 唐土人. Literally, this word means "a person from Tang (China)," but practically it was applied not only to the Chinese, but also to the Koreans, and even generally to all things foreign. And it was not necessarily a foreign origin that made something or somebody *morokoshibito*, it was the fact that "foreign" signified strange and incomprehensible. The foreign was novel, sophisticated, bizarre, and automata were all of those. Associations do not follow the rules of syllogism, and if the foreign was bizarre, and automata were

bizarre, then the conclusion was that there was something foreign in mechanical puppets that moved on their own.

3. Ihara Saikaku, *Dokugin hyakuin*, p. 38.
4. *Cha hakobi ningyō* 茶運人形.
5. In Japanese, the tea-drinking utensil is a bowl, *chawan* 茶碗. I refer to it as a "cup" because in English the word "bowl" evokes associations with a different kind of dining experience.
6. "Exquisite precision." *Seimyō* 精妙.
7. 平賀源内 (1728–1780).
8. Morishima Chūryō, *Kōmō zatsuwa*. The title can be translated in many ways. First of all, the word *kōmō* literally translates as "red hair"—a derogatory term coined upon the initial encounter of the Japanese with the Dutch. However, by the end of the eighteenth century the word lost its negative connotation, and often (as was the case with the previous title) was simply transliterated as "*oranda*." Second, the word *zatsu* could be elegantly translated as "miscellaneous." However, given the popular nature of the book, I prefer simpler words.
9. Katsuragawa Hoshū, 桂川 甫周 (1751–1809). The anatomical book in question is *The New Book of Anatomy* (*Kaitai Shinsho* 解体新書). The book is most often attributed to Sugita Genpaku, who in fact was only one of a large group of people who worked together on it.
10. In Europe too, clock-makers were the ones who built automata, employing their knowledge of time-measurement mechanisms to create life-like devices. See Voskhul, *Androids in the Enlightenment*.
11. Hosokawa, *Karakuri zui*.
12. 山路才助 (1761–1810).
13. 写天儀記. According to Motoyasu, *Kaga ni okeru karakuri zui shahon to gijutsu denpō*, p. 13.
14. Chuko Kongming (Jap. *Shokatsu Kōmei*) is Zhuge Liang (181–234), a legendary counselor immortalized in the ancient Chinese epic *The Three Kingdoms*. Zhuge Liang was known for his technical inventions and strategic planning. "Wooden oxen and running horses" (木牛流馬) was a military transportation cart, later described as a wheelbarrow.
15. Some modern-day historians suggested that the book was a professional manual for constructing clocks that disclosed professional secrets to the public. See Chaiklin, *Cultural Commerce and Dutch Commercial Culture*, p. 110. Timon Screech, *The Lens within the Heart*, p. 82. Nevertheless, as we shall see, no amateur could have learned how to build an automata by using this manual alone. Ryūji Yamaguchi, who examined the technical aspects of the text in detail, explicitly stated that it did not divulge secret knowledge to the public, but was, rather, a general introduction. Yamaguchi, *Nihon no Tokei*, pp. 267–68.
16. Martha Chaiklin, who investigated these reports, describes occasions at which Japanese officials borrowed clocks from the Dutch and later returned them in pieces. However, as seen in the previous sections, Japanese clock-makers were quite accustomed to dealing with Western clocks, so it is doubtful that they were not able to reassemble such clocks, even if they encountered a somewhat different design. It is more likely that these incidents involved amateurs, which would point to the fact that interest in clock mechanics was not confined to professionals. Chaiklin, *Cultural Commerce and Dutch Commercial Culture*, p. 94.
17. Hosokawa, *Karakuri zui*, p. 195.

18. Timon Screech claimed that clocks were interesting because they were mysterious. In "Clock Metaphors in Edo Period," p. 66, he writes: "The workings of clocks provided a metaphor for secret activity, or even surreptitious machinations. Clocks ran on privately and internally in a way that no amount of staring at their outsides could ever fathom." Given the design of the many clocks that exposed the inner working of the gears, I strongly disagree with this claim.
19. See, for example, *Tokei Seisaku sho, Tokei zatō no zu, Tokei zukai*. It is interesting to note that all of the above were written by anonymous authors, undated, and none are mentioned in the *Union Catalog of Early Japanese Books*. Another anonymous work, *Tokei zu*, preserved in Tokyo National Museum is the only one of the genre that appears in the catalog.
20. Anonymous, *Tokei zu*.
21. Uehara, *Tenmon rekigaku shoka shokanshū*, pp. 68–74.
22. 久米栄左衛門通賢 (1780–1841).
23. Cited in Suzuki Kazuyoshi et al., *"Kume Tsūken ni miru edo jidai no kagakugijutsu,"* p. 5.
24. Nakamura, *"Kume Eizaemon Tsūken no Nihon tenmon, sokuryōshijō niokeru ichi,"* pp. 9–10.
25. Baba Sajuro, *Tokei zusetsu*.
26. 司馬 江漢 (1737–1818).
27. Shiba, *Oranda Tensetsu*.
28. 普門円通 (1755–1834).
29. "Model" is *gi*, 儀. *Shumisengi mei narabini jo wakai*.
30. Ibid.
31. *Bukkoku rekishō hen*.
32. *Shumisen* 須弥山 in Japanese.
33. "Indian Calendar" *Bonreki* 梵暦. Written in this character, "Indian" means "related to Buddhism."
34. In his *History of Japanese Astronomy*, Shigeru Nakayama depicted Entsū as a representative of dogmatic forces holding onto the traditional theory (pp. 210–13). Nevertheless, his acceptance of existing methods of astronomical practice testifies to the fact that he was very much in line with the leading scholars of his time in terms of his approach to scientific investigation. We no longer discuss scientific developments in terms of paradigmatic shifts, but if we were to do so today, it would be fair to say that Entsū's example shows that differentiation along paradigmatic lines should not be based on theoretical differences but rather on practical approaches and underlying assumptions.
35. Inō Tadataka, *Bukkokureki shōhen sekimō*.
36. In "Vision and Reality," Okada Masahiko provides an excellent analysis of Entsū's work.
37. Entsū, *Shumisengi mei narabini jo wakai*.
38. 田中久重 (1799–1881), からくり儀右衛門.
39. In Japanese the puppet is called *Arrow Shooting Child* 弓引き童子, which in itself is gender-less. Most of the present-day viewers assume that the shooter is a boy, but Hisashige's notes clearly show that he referred to the puppet as 女子 — a girl. *Shoki kōan zu*.
40. Asahina Teiichi and Oda Sachiko, "'Myriad-Year Clock' Made by G. H. Tanaka 100 Years Ago in Japan," pp. 2–3.
41. Seiko Museum lists their *Model of Sumeru World* as dating back to 1847.
42. *Shumisengi* 須弥山儀.

43. According to Sawada Taira, there are seven surviving *Shumisegi* in Japan. Sawada Taira, "*Shumisengi monogatari*," pp. 17–23.
44. 佐田介石(1818–1882).
45. *Shijitsutōshōgi* 視実等象儀.
46. The astronomy practiced in Japan at the time was heavily based on Western astronomical theory and practice but incorporated several major elements of Chinese astronomy. Moreover, Entsū objected not only to the geokinetic view but also to the mere idea that the world is spherical, and hence rejected Chinese astronomical tradition as well.
47. Takehiko Hashimoto, "Tanaka Hisashige and His Myriad Year Clock," pp. 42–44.
48. 広瀬元恭 (1821–1870).
49. *Man-nen dokei* 萬年時計.
50. Several modern investigations agreed that the clock could work for as many as 225 days straight. Asahina and Oda, "'Myriad-Year Clock' Made by G. H. Tanaka 100 Years Ago in Japan," p. 3; Suzuki Kazuyoshi et al., "Mechanism of '*Man-non dokei*,'" p. 3.
51. For sexagenarian cycle, see appendix 4.
52. Hashimoto, "Tanaka Hisashige and His Myriad Year Clock," pp. 31–45.
53. Suzuki et al., "Mechanism of '*Man-non dokei*,'" pp. 2–4.
54. Ibid.
55. For discussion of the image of Western numbers, see Annick Horiuchi, "Honda Toshiaki (1743–1820) *ou l'Occident comme utopie*," p. 411–48. A specific reference linking number counting and Western watches can be seen in Tadano Makuzu's "*Hitori Kangae*"—see Tadano Makuzu, "Solitary Thoughts: A Translation of Tadano Makuzu's *Hitori Kangae* (2)," trans. Goodwin et al., p. 193.
56. Kyūshidō Shujin 求巳堂主人 (according to Watanabe Toshio, this was one of the pen names of Shibukawa Kagesuke 渋川景佑), *Shōchū jishingi jimō* 掌中時辰儀示蒙 (Pointing out the ignorance about pocket watches), 文政六.
57. Fujimura Heizō 藤村平三, *Jimeisho jiban kō* 自鳴鐘時盤考 (Investigation of the clock dial), 文政六. Published by clock-maker Tōda Tozaburō 戸田東三郎.
58. Shūyūdō Shujin, *Waei tsūin iroha benran*.
59. Kawaguchi Jusai, *Sanshin Hatsuei*.
60. Anonymous, *Jishingi Teiyō*.
61. See also Yanagiwa Shunsan, *Seiyō tokei benran*, Anonymous, *Wayō taishō tokei hyō*.
62. Ogawa Tomotada, *Seiyō jishingi teikoku kassoku*.
63. Ibid.
64. Around the second half of February, according to the current calendar.
65. Ogawa Tomotada, *Seiyō jishingi teikoku kassoku*.
66. Ibid.

CHAPTER EIGHT

1. Matsudaira Sadanobu 松平定信 (1759–1829); Nakai Chikuzan中井竹山 (1730–1804).
2. Calendrical divinational indicators: *rekichū* 暦柱.
3. For divinational practices in calendars see Leinss, "*Japanische Lunisolarkalender der Jahre Jōkyō 2 (1685) bis Meiji 6 (1873). Zeicheninventar.*"
4. "Ignorant Fables." *Bōtan* 妄誕. *Sōbōkigen*. Cited in Okada Yoshirō, *Meiji kaireki*, p. 37.
5. The combination of these three words (天文、地理、渡海) appears on every page

of the first volume of *The Tales of the West*, encouraging different ways of "developing" (開く) these fields. Honda Toshiaki, *Seiiki monogatari*, pp. 87–162.

6. *Seiiki monogatari*, p. 106–7.
7. *Ahō* 阿房. *Seiiki monogatari*, p. 109.
8. 中井履軒 (1732–1817).
9. *Kashoreki*, pp. 39–40.
10. 山片蟠桃 (1748–1821).
11. Yamagata, *Yume no shiro*.
12. 塚本明毅 (1833–1885).
13. "率ネ妄誕不稽ニ属シ民知ノ開達ヲ妨ルモノ少シトセス" in *Tsukamoto Aketake Kengi*, pp. 126–28.
14. "率ネ妄誕無稽ニ属シ人智ノ開達ヲ妨ルモノ少シトセス" in Imperial House, *Shōsho*, p. 117.
15. Inconvenience: *fuben* 不便.
16. 市川齋宮 (1818–1899).
17. *Rekihō gian*, p. 73.
18. The condescending attitudes of Japanese intellectuals were not exceptional in the world. Half a century later Arabic intellectuals would express similar attitudes toward the way "the people" treated time. See Vanessa Ogle, *The Global Transformation of Time*, and On Barak, *On Time*.
19. *Rekihō gian*, p. 74.
20. Okada, *Meiji kaireki*, pp. 79–80.
21. *Kairekiben* 改暦弁.
22. 福沢諭吉 (1835–1901).
23. *Fukuzawa Yukichi zenshū*, p. 324.
24. Ibid.
25. Ibid.
26. 無学文盲の馬鹿者なり. Ibid., p. 325.
27. Ibid.
28. In "Teaching Punctuality," Nishimoto Ikuko discusses the introduction of the "Time Is Money" phrase into the Meiji educational system.
29. Vanessa Ogle, *The Global Transfomation of Time*.
30. Cited in Okada, *Meiji kaireki*, p. 78.
31. *Tasshi* concerning the reform. Cited in Okada Yoshirō, *Meiji kaireki*, p. 119.
32. Tsukamoto Aketake Proposal, cited in Okada Yoshirō, *Meiji kaireki*, p. 127.
33. *Kairekiben*, pp. 329–30.
34. Ichikawa Saigū, in Okada Yoshirō, *Meiji kaireki*, p. 74.
35. For detailed discussion see chapter 2.
36. See chapter 3.
37. Takahashi Tamagusuku, *Kaireki subeki koto* 改暦スベキ事, and *Kōkai no justu wo manabashimu koto* 航海ノ術ヲ学バシムベキ事; Hirokawa Seiken, *Rekihō kaikaku no koto* 暦法改革之事, and *Chiri kyūmutaru koto* 地理急務タル事. See Okada Yoshirō, *Meiji kaireki*, p. 78.
38. 学問のススメ and 訓蒙窮理圖解 accordingly.
39. Ichikawa Saigū, in Okada Yoshirō, *Meiji kaireki*, p. 74.
40. Kumagai Tōshū, *Jishingi no setsu*.
41. Fukuzawa Yukichi, "Preface to the collected works of Fukuzawa," in *The Autobiography of Fukuzawa Yukichi*, trans. Eiichi Kiooka (Tokyo: Hokusaido Press, 1981), pp. 76–77. Emphasis added by the author.

42. Marius Jansen, *The Making of Modern Japan* (Belknap Press, 2002), 460.
43. There is a vast amount of literature concerning implementation of the reform and people's response to it. In Japanese context particularly notable are Hashimoto and Kuriyama, eds., *Chikoku no Tanjō* and Nishimoto's *Jikan ishiki no kindai*. Vanessa Ogle shows that the implementation of "modern time" was not straightforward in Europe either. See *The Global Transformation of Time.*
44. I discuss the process of learning new habits of measuring time with Western watches in "Translating Time: Habits of Western Style Timekeeping in Late Tokugawa Japan."

BIBLIOGRAPHY

ABBREVIATIONS

Gakushiin—Japanese Academy of Science (Gakushiin) Archive
Hazama—Hazama Bunko Archive
Kanō—Tohoku University Kanō Archive
NAOJ—National Astronomical Observatory of Japan, Mitaka Library
NDL—National Diet Library Rare Books Collection
NST—*Nihon Shisō Taikei* 日本思想大系 [*Compendium of Japanese Thought*]
TNM—Tokyo National Museum Archive
Wasan—Tohoku University Wasan (Japanese Mathematics) Database
Waseda—Waseda University Kotenseki Sogo Database of Japanese and Chinese Classics

SOURCES

Primary sources are given in transliteration, original, and translation.

Abe Yasukuni 安倍泰邦. *Rekihō shinsho* 暦法新書 [New book on calendrical methods]. 1754. NDL.

Adas, Michael. *Machines as the Measure of Men.* Ithaca: Cornell University Press, 2015.

Alder, Ken. *The Measure of All Things.* New York: Free Press, 2002.

Alpers, Svetlana. *The Art of Describing.* Chicago: University of Chicago Press, 1983.

Andrews, William, ed. *The Quest for Longitude.* Cambridge, MA: Collection of Historical Scientific Instruments, Harvard University, 1996.

Anonymous. *Jūniji shōko no kazu* 十二時鐘鼓之数 [The numbers of the twelve hours of bells and drums]. 1720. TNM.

———. *Tokei zu* 時計図 [Illustrations of clocks]. Date unknown. TNM.

———. *Tokei seisaku sho* 時計製作書 [Manual for making clocks]. Date unknown. Kanō.

———. *Tokei zatō no zu* 時圭座等之図 [Illustrations of clock positioning]. Date unknown. Kojū Bunko.

———. *Tokei zukai* 時計図解 [Illustrated explanation of clocks]. NDL.

———. *Jishingi Teiyō* 時辰儀提要 [An outline of Western watches]. Date unknown. Gakushiin.

———. *Wayō taishō tokei hyō* 和洋対照時計標 [A comparison table of Japanese and Western clocks]. Date unknown. NDL.

Asai Tadashi 浅井忠. *Suiyōkyūgi: wadokei chōsa hōkoku* 垂揺球儀: 和時計調査報告. Otsu: Wadokei Gakkai, 1983.

Asahina Teiichi and Oda Sachiko. "'Myriad-Year Clock' Made by G. H. Tanaka 100 Years Ago in Japan." *Bulletin of National Science Museum* 1, no. 2 (September 1954): 1–21.

Asao Naohiro 朝尾直弘. "*Jidai kubunron*." In *Nihon Tsūshi*, 97–122. Tokyo: Iwanami Shoten, 1995.

Araki Toshima 荒木俊馬. *Jisoku to Rekihō* 時側と暦法. Tōkyō: Kōseisha, 1963.

———. *Toki to koyomi* 時と暦. Tokyo: Haneda Shobō, 1942.

Arnheim, Rudolf. *Visual Thinking*. Berkeley: University of California Press, 1969.

Arisaka Takamichi 有坂隆道. *Nihon yōgakushi no kenkyū* 日本洋学史の研究. Osaka: Sōgensha, 1968.

Aubin, David, Charlotte Bigg, and H. Otto Sibum, eds. *The Heavens on Earth*. Durham: Duke University Press, 2010.

Baba, Sajūrō 馬場佐十郎. *Tokei zusetsu* 時計図説[Illustrated explanation of clocks]. Wasan.

Baird, Davis. *Thing Knowledge*. Berkeley: University of California Press, 2004.

Bashō Matsuo 松尾芭蕉. *Oku no hosomichi* おくの細道 [The narrow road to the interior]. Tokyo: Iwanami Bunko, 2004. First published in 1702.

Barak, On. *On Time: Technology and Temporality in Modern Egypt*. Berkeley: University of California Press, 2013.

Bedini, Silvio. *The Scent of Time*. Philadelphia: American Philosophical Society, 1963.

———. "Oriental Concepts of the Measure of Time." In *The Study of Time II*, edited by J. T. Fraser, N. Lawrence, and D. Park, 451–84. Berlin: Springer-Verlag, 1975.

———. *The Trail of Time—Shih-chien ti tsu-chi: Time Measurement with Incense in East Asia*. Cambridge: Cambridge University Press, 1994.

Berry, Elizabeth. *Japan in Print*. Berkeley: University of California Press, 2006.

Beukers, Harmen. "Dodonaeus in Japanese: Deshima Surgeons as Mediators." In *Dodonaeus in Japan*, edited by Willy vande Walle and Kazuhiko Kasaya, 281–98. Leuven: Leuven University Press, 2001.

Bijker, Wiebe E., and Trevor J. Pinch. "The Social Construction of Facts and Artifacts: Or How the Sociology of Science and the Sociology of Technology Might Benefit Each Other." In *The Social Construction of Technological Systems*, edited by Bijker et al., 17–50. Cambridge, MA: MIT Press, 1987.

Birth, Kevin. *Objects of Time*. New York: Palgrave McMillan, 2012.

Bordieu, Pierre. *Outline of a Theory of Practice*. Cambridge: Cambridge University Press, 1977.

Bowers, John. *When the Twain Meet*. Baltimore: Johns Hopkins University Press, 1980.

Bowker, Geoffrey C. *Memory Practices in the Sciences*. Cambridge, MA: MIT Press, 2005.

Brandes, Wilhelm. "The Secret of Japanese Clock Dial." *NAWCC Bulletin*, no. 172 (October 1974): 531–65.

Brown, Philip. *Cultivating Commons: Joint Ownership of Arable Land in Early Modern Japan*. Honolulu: University of Hawaii Press, 2011.

———. "Never the Twain Shall Meet: European Land Survey Techniques in Tokugawa Japan." *Chinese Science* 9 (1989): 53–79.

Burnett, Graham. *Masters of All They Surveyed*. Chicago: University of Chicago Press, 2000.

Buys, Egbert. *Nieuw en volkomen Woordenboek van Konsten en Weetenschappen* [A new and complete dictionary of terms of arts and sciences]. Amsterdam: S. J. Baalde, 1769–1778.

Canales, Jimena. *A Tenth of a Second*. Chicago: University of Chicago Press, 2009.

———. *The Physicist and the Philosopher*. New Jersey: Princeton University Press, 2016.

Chaiklin, Martha. *Cultural Commerce and Dutch Commercial Culture*. Leiden: CNWS, 2003.

Clancey, Greg. *Earthquake Nation*. Berkeley: University of California Press, 2006.

Collins, Harry M. *Tacit and Explicit Knowledge*. Chicago: University of Chicago Press, 2010.

———. "The TEA Set: Tacit Knowledge and Scientific Networks." In *The Science Studies Reader*, edited by Mario Biagoli, 95–109. New York: Routledge, 1999.

Cook, Harold. *Matters of Exchange*. New Haven: Yale University Press, 2007.

Cooper, Michael. *They Came to Japan*. Berkeley: University of California Press, 1965.

Crary, Jonathan. *Techniques of the Observer*. Cambridge, MA: MIT Press, 1992.

Cryns, Frederick. *Edo jidai ni okeru kikaironteki shintaikan no juyō*. Kyoto: Rinsen Shoten, 2006.

Cullen, Christopher. *Astronomy and Mathematics in Ancient China: The Zhou pi suan jing*. Cambridge: Cambridge University Press, 1995.

Daston, Lorraine. *Biographies of Scientific Objects*. Chicago: University of Chicago Press, 2000.

———. "On Scientific Observation." *Isis* 99, no. 1 (2008): 97–110.

———. *Things That Talk*. Cambridge, MA: MIT Press, 2004.

Dohrn-van Rossum, Gerhard. *History of the Hour*. Translated by Thomas Dunlap. Chicago: University of Chicago Press, 1996.

Dolce, Lucia. *The Worship of Stars in Japanese Religious Practice*. Bristol, UK: Culture and Cosmos, 2007.

Edgerton, David. *Shock of the Old*. Oxford: Oxford University Press, 2006.

Edo no Monozukuri International Symposium. *Kinsei Kagaku gijutsu no DNA to gendai haiteku ni okeru wagakuni kagaku gijutsu no aidentiti no kakui*. Kyoto: Ministry of Education target project *Edo no Monozukuri*, 8th International Symposium, 2007.

Edwardes, Ernest L. *Weight-Driven Dutch Clocks and Their Japanese Connections*. Ashbourne: Mayfield Books in association with the Antiquarian Horological Society, 1996.

Elman, Benjamin. *On Their Own Terms*. Cambridge, MA: Harvard University Press, 2005.

Endō Katsumi 遠藤克己. *Kinsei onmyōdōshi no kenkyū*. Hino: Mirai Kōbō, 1985.

Endō Motoo 遠藤元男. *Shokunin no seiki, kinsei hen*. Tokyo: Yūzankaku, 1991.

Endō Ōsekishi 遠藤黄赤子. *Chūya chōtan no zu* 昼夜長短之図 [Illustration of long and short days and nights].1709. Kanō.

Endō Takanori 遠藤高璟. *Kanazawa Jishōki* 金沢時鐘記 [The record of bell-keeping in Kanazawa]. 1823. Kōju Bunko.

———. *Kenjō onjiki yurai narabini yōhō no oboe* 献上之御時規由来並用法之覚 [Memoir of the presented clock and its use]. 1849. Gakushiin.

———. *Shahō shinjutsu* 写法新術 [New art of depiction].1850. Gakushiin.

———. *Takezawa goten soku koku go kibutsu yōhō* 竹沢御殿測刻御器物用法 [The method of use of time measurement devices of Lord Takezawa]. 1823. Kojū Bunko.

———. *Tokei yōhō ki* 時規用法記 [Manual for using the clock]. 1794. Tohoku University Archive.

Engelfriet, Peter M. *Euclid in China*. Leiden: Brill, 1998.

Evans, James. *The History and Practice of Ancient Astronomy*. New York: Oxford University Press, 1998.

Ferguson, Eugene S. *Engineering and the Mind's Eye*. Cambridge, MA: MIT Press, 1992.

Fraser, J.T. *The Voices of Time*. New York: G. Braziller, 1966.

Frumer, Yulia. "Before Words: Reading Western Astronomical Texts in Early 19th Century Japan." *Annals of Science* 73, no. 2 (2016): 170–94.

———. "Translating Time: Habits of Western Style Timekeeping in Late Tokugawa Japan." *Technology and Culture* 55, no. 4 (October 2014): 785–820.

Fujimura, Heizō 藤村平三. *Jimeisho jiban kō* 自鳴鐘時盤考 [Investigation of the clock dial]. 1823. NDL.

Fukada Kōjitsu 深田香実 *et al. Owari shi* 尾張志 *[The History of Owari Domain]. 1843. Naikaku Bunko.*

Fukuoka Maki. *The Premise of Fidelity*. California: Stanford University Press, 2012.

Fukuzawa Yukichi 福沢諭吉. *Kairekiben* 改暦弁 [A treatise on the calendrical reform]. Vol. 3 of *Fukuzawa Yukichi zenshū*. Tokyo: Iwanami Shoten, 1969.

———. "Preface to the Collected Works of Fukuzawa." In *The Autobiography of Fukuzawa Yukichi*, translated by Eiichi Kiooka. Tokyo: Hokusaido Press, 1981.

Fumon Entsū 普門円通. *Bukkoku rekishō hen* 仏国暦象編 [Book of astronomical phenomena in Buddhist countries]. 1810. Kanō.

———. *Shumisengi mei narabini jo wakai* 須弥山儀銘並序和解 [Japanese explanation of model of Mt. Sumeru]. 1813. Kanō.

Galison, Peter. *Einstein's Clocks, Poincaré's Maps*. London: W. W. Norton, 2003.

———. *Image and Logic*. Chicago: University of Chicago Press, 1997.

———. "Trading Zone: Coordinating Action and Belief." In *The Science Studies Reader*, edited by Mario Biagioli, 137–60. New York: Routledge, 1999.

Gardner, Nakamura Ellen. *Practical Pursuits*. Cambridge, MA: Harvard University Press, 2005.

Gell, Alfred. *The Anthropology of Time*. Oxford: Berg, 1992.

Golovnin, Vasiliy. *Zapiski flota kapitana Golovnina o prikliyucheniiakh ego v plenu u iapontsev* [The diaries of the captain of the Russian fleet, Golovnin, about his adventures in Japanese captivity]. Khabarovsk: Khabarovskoe knizhnoe izd-vo, 1972.

Golvers, Noël. *Ferdinand Verbiest, S.J. [1623–1688] and the Chinese Heaven*. Leuven: Leuven University Press, 2003.

Gombrich, E. H. *Art and Illusion*. New York: Pantheon Books, 1960.

Goodman, Grant K. *Japan and the Dutch 1600–1853*. Richmond, UK: Curzon Press, 2000.

Gotō Godōan 後藤梧桐庵. *Oranda banashi* 紅毛談 [Dutch tales]. 1765. Waseda.

Grafton, Anthony, and Daniel Rosenberg. *Cartographies of Time*. Princeton: Princeton Architectural Press, 2010.

Gurevich, Aaron J. "Time as a Problem of Cultural History." In *Cultures and Time*, edited by L. Gardet. Paris: Unesco, 1976.

Harding, Sandra, ed. *Postcolonial Science and Technology Studies Reader*. North Carolina: Duke University Press, 2011.

Hashimoto Manpei 橋本万平. *"Jikan ni genkakuna Hidetada." Wadokei* 2 (October 1993): 1–2.

———. *Nihon no jikoku seido*. Tokyo: Kosensho, 1957.

Hashimoto Takehiko 橋本毅彦. "The Adoption and Adaptation of Mechanical Clocks in Japan." In *Science between Europe and Asia: Historical Studies on the Transmission, Adoption, and Adaptation of Knowledge*, edited by Feza Günergun and Dhruv Raina, 137–49. Dordrecht: Springer, 2011.

———. *Historical Essays on Japanese Technology*. Tokyo: Collection UTCP, 2009.

———. "Mechanization of Time and Calendar: Tanaka Hisashige's Myriad Year Clock and Cosmological Model." *UTCP Bulletin* 6 (2006): 47–55.

———. *"Tokei no gijutsuteki tokuchō to shakaiteki igi ni kansuru rekishiteki kenkyū."* Report of the Grant-in-Aid for Scientific Research (no.14023202) by Ministry of Education, Science, Sports and Culture. In *Edo no Monozukuri*. Tokyo: Monbukagakusho, 2006.

———. *"Toki o hakaru—wadokei no shinka to Edo no jikoku seido."* In *Edo no Monozukuri*, 40–44. Tokyo: Monbukagakusho, 2004.

———. *"Wadokei no seido wo megutte." Bulletin of the National Science Museum* 414 (2003): 9–11.

Hashimoto Takehiko 橋本毅彦, and Shigehisa Kuriyama 栗山茂久, eds. "The Birth of Tardiness," special issue, *Japan Review* no. 14 (2002).

———. *Chikoku no Tanjō*. Tokyo: Sangensha, 2001.

Hattori Yoshitaka 服部義高. *Kaisen anjō roku* 廻船安乗録 [Record of safe seafaring]. 1813. NDL.

Hayashi Makoto 林淳. *"Igo to tenmon."* In *Bunkashi no shosō*, edited by Ōsumi, 258–87. Tokyo: Yoshikawa Kōbunkan, 2003.

———. *Kinsei Onmyōdo no Kenkyū*. Tokyo: Yoshikawa Kōbunkan, 2005.

———. *Tenmongata to onmyōdō*. Tokyo: Yamakawa Publishing, 2006.

———. *"Tōhoku no Onmyōdō to Tenmongaku." Aichi Gakuin daigaku ningen bunka kenkyū johō* 20 (2005): 39–51.

Hayashi Makoto and Matthias Hayek, eds. *"Onmyōdō* in Japanese History." *Japanese Journal of Religious Studies* 40, no. 1 (2013): 1–18.

Hayek, Matthias. "The Eight Trigrams and Their Changes: An Inquiry into Japanese Early Modern Divination." *Japanese Journal of Religious Studies* 38, no. 2 (2011): 329–68.

———. *"Les Manuels de divination japonais au début de L'époque d'Edo (xviie siècle): décloisonnement, compilation, et vulgarisation." Extrême-Orient Extrême-Occident* 35 (2013): 83–112.

Hazama Shigetomi 間重富. *Tenchi nikyūyōhō kihyōsetsu* 天地二球用法記評説 [Criticism of "The Use of Celestial and Terrestrial Globe"]. 1798. Gakushiin.

———. *Suikyūseigi* 垂球精義 [Detailed exposition of suspended disk]. 1804. Gakushiin.

Hazama Shigeyori 間重新. *Senkō taigyō sensei jiseki ryakki* 先考大業先生事迹略記 [Abbreviated record of work of the great master, my late father]. In *Tenmonrekigakushijō ni okeru Hazama Shigetomi to sono ikka*, edited by Watanabe Toshio, 448–76. Tokyo: Yamaguchi Shoten, 1943.

Hendry, Joy. "Cycles, Seasons and Stages of Life." In *The Story of Time*, edited by Kristen Lippincott. London: Merrell Holberton Publishers, 1999.

Hirai Sumio 平井澄夫. *"Taiyō no hari." Wadokei Kenkyū* 1 (June 1992): 30–31.

———. *Tokei no hanashi*. Tokyo: Asahi Shinbun Shuppan, 2001.

Hiraoka Ryuji. "The Transmission of Western Cosmology to 16th Century Japan." In *The Jesuits*, edited by Luís Saraiva and Catherine Jami. Hackensack, NJ: World Scientific, 2008.

Hirose Hideo 広瀬秀雄. *Koyomi*. Tokyo: Kondō Shuppansha, 1984.

Ho Peng Yoke. *Chinese Mathematical Astrology: Reaching Out to the Stars*. London: Routledge/Curzon, 2003.

Honda Toshiaki 本多利明. *Kōekiron* 交易論 [Treatise on commerce]. 1801. In NST, vol. 44, 175–83.

———. *Seiiki monogatari* 西域物語 [Tales of the West]. 1798. In NST, vol. 44, 87–162.

———. *Shisei chiji kagami* 視星知時鑑 [Looking at the stars, knowing the hours]. 1789. Kanō.

———. *Tokai shinpō* 渡海新法 [New method of crossing the sea]. 1804. Kanō.

———. *Toki no kane zuke* 時の鐘附 [Appendix to Time-Bells]. 1816. Kanō.

Horiuchi Annick. "Honda Toshiaki (1743–1820) *ou l'Occident comme utopie."* In *Repenser l'ordre, repenser l'héritage*, edited by F. Girard, A. Horiuchi, and M. Macé, 411–48. Paris: Droz, 2002.

———. *Japanese Mathematics in the Edo Period (1600–1868)*. Basel: Birkhäuser, 2010.

———. *"La Recomposition du Paysage Mathematique Japonais."* In *L'Europe Mathématique:*

Histoires, Mythes, Identités, edited by Catherine Goldstein, Jeremy Gray, and Jim Ritter, 248–68. Paris: Editions de la Maison des sciences de l'homme, 1996.

Hosokawa Hanzō 細川半蔵. *Karakuri zui* 機巧図彙 [Illustrated manual of curious machines]. 1796. NDL.

Hotta Ryōhei 堀田両平. *Tokei no bunken mokuroku*. Tokyo: Hotta Ryohei, 1971.

———. *Tokei to koyomi no sashie mokuroku*. Tokyo: Hotta Ryohei, 1971.

Howell, David. *Capitalism from Within*. Berkeley: University of California Press, 1995.

———. *Geographies of Identity in Nineteenth-Century Japan*. Berkley: University of California Press, 2005.

Hoyanagi Mutsumi 保柳睦美. *Ino Tadataka no kagakuteki gyōseki*. Tokyo: Kokin Shoin, 1974.

Hubert, Henri. *Essay on Time: A Brief Study of the Representation of Time in Religion and Magic*. Translated by Robert Parkin. Oxford: Durkheim Press, 1999.

Ichikawa Saigū 市川齋宮. *Rekihō gian* 暦法議案 [Proposal concerning calendrical method], cited in Okada, *Meiji kaireki*. Tokyo: Taishukan Shoten, 1994.

Ihara Saikaku 井原西鶴. *Dokugin hyakuin* 独吟百韻 [Hundred rhymes recited in solitude]. In *Zusetsu karakuri: asobi no hyakka.*, edited by Tatsukawa Shōji et al. Tokyo: Kawade Shobo Shinsha, 2002.

———. *Kōshoku ichidai otoko* 好色一代男 [The life of amorous man].1682. Tokyo: Iwanami Shoten, 2015.

———. *Nippon Eitaigura* 日本永代蔵 [Eternal storehouse of Japan]. 1688. Tokyo: Kadokawa Gakugei Shuppan, 2009.

Iida Yoshirō 飯田芳郎. *"Kōkai yōhō setsuyaku nitsuite shōkai." Kaijishi kenkyū* 25 (1975): 112–29.

———. *Nihon kōkaijutsushi : kodai kara bakumatsu made*. Tokyo: Hara Shobō, 1980.

———. *"Tokai shinpō ni okeru kōkaigaku." Kaijishi* 28 (1977): 1–25.

Ikeda Richiko, and Zheng Wei. "An Analysis of Chinese and Japanese Concepts of Time Seen through the History of Mechanical Clocks." *Human Communication Studies* 30 (2002): 1–21.

Imperial House. *Shōsho* 詔書 [Imperial restrict], cited in Okada, *Meiji kaireki*, 117. Tokyo: Taishukan Shoten, 1994.

Inō Tadataka 伊能忠敬. *Bukkokureki shōhen sekimō* 仏国暦象編斥妄 [The backwardness of the Book of Astronomical Phenomena in Buddhist countries]. Date unknown. Waseda.

———. *Kokugun chūya jikoku* 国郡昼夜時刻稿本 [The times of days and nights in provinces and districts]. Date unknown. Gakushiin.

———. *Nihon keiido jissoku* 日本経緯度実測 [Measurements of longitudes and latitudes of Japan]. 1800. Kanō.

———. *Sokuryō nikki* 測量日記 [Surveying diary]. Original undated. Reprint edited by Sakuma. Tokyo: Ōzorasha, 1998.

Ishii Kendo 石井研堂, ed. *Ikoku hyōryu kidanshu*. Tokyo: Shin Jinbutsu Ōraisha, 1976.

———. *Nihon kaijishi no shomondai* 日本海事史の諸問題 [Various problems in the history of Japanese navigation]. Tōkyō: Bunken Shuppan, 1995.

Ishikawa ken. *Ishikawa kenshi shiryō*. Kinsei hen. Kanazawa: Ishikawa-ken, 2000.

Jami, Catherine, and Luís Saraiva, eds. *The Jesuits: The Padroado and East Asian Science (1552–1773)*. Hackensack, NJ: World Scientific, 2008.

Jansen, Marius. *The Making of Modern Japan*. Cambridge, MA: Belknap Press of Harvard University Press, 2000.

Juznic, Stanislav. "Chinese-Slovenian Astronomer Hallerstein: Provincial and Visitor to Japanese Jesuit Province." *Historia Scientarum* 19, no. 1 (July 2009): 55–71.

Kaempfer, Engelbert. *The History of Japan*. Translated by J. G. Scheuchzer. Glasgow: J. MacLehose and sons, 1906.

Kagoshimaken shiryō kankō iinkai. *"Urashima sokuryō no zu."* Vol. 10 of *Kagoshima Shiryōshū*. Kagoshima: Kagoshimakenritsu toshokan, 1970.

Kanda Shigeru 神田茂. *Nihon Tenmon Shiryou*. Tokyo: Hara Shobō, 1978.

Kawaguchi Jusai 川口戌斎. *Sanshin Hatsuei* 三針発映 [*On the movement of the three hands*]. 1867. Kanō.

Kawasaki, Michio 河崎倫代. *Tenmon rekigakusha Nishimura Tachū den: Echū Jōhana no hito*. Jōhanamachi: Nishimura Tachū kinen hikyousankai, 2001.

Kinoshita Yasuhiro 木下泰宏.*"Mannendokei no bunkai chōsa kara mita Bakumatsu seiyō gijutsu no juyō nikansuru kenkyū."* PhD diss., Tokyo Technical Institute, 2015.

Kondō Katsuno 近藤勝之. *"Furiko en gurafu shiki mojiban kakedokei." Wadokei Kenkyū* 38 (December 2003): 4–8.

———. *"Saishoki makuradokei ni tsuite." Wadokei Kenkyū* 41 (January 2007): 1–13.

———. *"Wadokei to gendaijikoku no sai ni tsuite." Wadokei Kenkyū* 41 (January 2007): 14–16.

Kornicki, Peter. "Literacy Revised: Some Reflections on Richard Rubinger's Findings." *Monumenta Nipponica* 56, no. 3 (Autumn 2001): 381–95.

Koselleck, Reinhart. *Futures Past: On the Semantics of Historical Time*. Translated by Keith Tribe. Cambridge, MA: MIT Press, 1985.

Kusa Dōin 久佐道允, *Shichijūnikō-shō* 七十二候抄 [Excerpt of the Seventy Two Seasons]. Vol. 1–2. 1686. Waseda.

Kumagai Tōshū 熊谷東洲. *Jishingi no setsu* 時辰儀の説 [Explanation of a clock]. 1877. Wasan.

Kruzenshtern, Ivan Fedorovich. *Voyage around the world in the years 1803, 1804, 1805*. Vol. 1. London: Printed by C. Roworth for J. Murray, 1813.

Kyūshidō Shūjin 求巳堂主人. *Shōchū jishingijimō* 掌中時辰儀示蒙 [Pointing out mistakes about the hand watch]. 1860 copy. Tokyo University Archive. First published in 1823.

Landes, David. *Revolution in Time*. Cambridge, MA: Belknap Press of Harvard University Press, 1983.

La Pérouse, Jean-François de Galaup. *The voyage of La Pérouse round the world*, London: Printed for John Stockdale, 1798.

Latour, Bruno. "Do Scientific Objects Have a History? Pasteur and Whitehead in a Bath of Lactic Acid." *Common Knowledge* 5, no. 1 (1996): 76–91.

———. "Drawing Things Together." In *Representation in Scientific Practice*, edited by M. Lynch and S. Woolgar, 19–68. Cambridge, MA: MIT Press, 1990.

———. *Laboratory Life*. Beverly Hills, CA: Sage Publications, 1979.

———. "Mixing Humans with Non-humans: Sociology of a Door-Closer. " *Social Problems* 35 (1988): 298–310.

———. *Science in Action*. Cambridge, MA: Harvard University Press, 1987.

———. *We Have Never Been Modern*. Hemel Hempstead: Harvester Wheatsheaf, 1993.

Le Goff, Jacques. *Time, Work, and Culture in the Middle Ages*. Translated by Goldhammer. Chicago: University of Chicago Press, 1980.

Lehoux, Daryn. *Astronomy, Weather, and Calendars in the Ancient World*. Cambridge: Cambridge University Press, 2007.

Leinss, Gerhard. *"Japanische Lunisolarkalender der Jahre Jōkyō 2 (1685) bis Meiji 6 (1873). Aufbau und inhaltliche Bestandsaufnahme." Japonica Humboldtiana* 10 (2006): 5–89.

———. *"Japanische Lunisolarkalender der Jahre Jōkyō 2 (1685) bis Meiji 6 (1873). Zeicheninventar." Japonica Humboldtiana* 11 (2007): 53–78.

Levin, Iris, and Dan Zakay, eds. *Time and Human Cognition*. Amsterdam: Elsevier Science, 1989.

Liao, Pin. *Clocks and Watches of the Qing Dynasty*. Beijing: Foreign Languages Press, 2002.

Lippincott, Kristen. *The Story of Time*. London: Merrell Holberton Publishers, 1999.

Lynch, Michael. "Extending Wittgenstein: The Pivotal Move from Epistemology to the Sociology of Science." In *Science as Practice and Culture*, edited by Andrew Pickering, 215–65. Chicago: University of Chicago Press, 1992.

Lynch, Michael, and John Law. "Pictures, Texts, and Objects." In *The Science Studies Reader*, edited by Mario Biagioli, 317–41. New York: Routledge, 1999.

Mahoney, Michael S. "Christian Huygens: The Measurement of Time and of Longitude at Sea." In *Studies on Christiaan Huygens*, edited by H. J. M. Bos et al., 234–70. Lisse: Swets, 1980.

———. "Reading a Machine" http://www.princeton.edu/~hos/h398/readmach/modelt.html.

Makieshi Genzaburō 蒔絵師源三郎. *Jinrin kinmōzui*人倫訓蒙図彙 [Encyclopedia of professions]. Book 5, *Artisans*. 1690. NDL.

Mamiya Rinzō 間宮林蔵, and Murakami Teisuke 村上貞助. *Kitaezo zusetsu* 北蝦夷図説 [Illustrated explanations of Northern Ezo]. 1807. Waseda.

Mao Tokiharu 万尾時春. *Kikubuntōshū* 規矩分等集 [Compendium on (the art of) compass, ruler, and division]. 1722. NDL.

Marcon, Federico. *The Knowledge of Nature and the Nature of Knowledge in Early Modern Japan*. Chicago: University of Chicago Press, 2015.

Maruyama Masao. *Studies in the Intellectual History of Tokugawa Japan*. Tokyo: University of Tokyo Press, 1974.

Matsuzaki Toshio. *Edo Jidai no Sokuryōjustu*. Tokyo: Sōgokagakushuppan, 1979.

McCrossen, Alexis. *Marking Modern Times*. Chicago: University of Chicago Press, 2013.

McGee, Dylan. "Turrets of Time: Clocks and Early Configuration of Chronometric Time in Edo Fiction." *Early Modern Japan* 19 (2011): 44–57.

Meiji Government. *Tasshi* 達 [Government notice], cited in Okada, *Meiji kaireki*, 119. Tokyo: Taishukan Shoten, 1994.

Minamoto Sanuyoshi 源誠美. *Tenji Meikai* 天時明解 [Explication of heavenly time]. 1802. Kanō.

Ministry of Justice Research Dept., ed. *Tokugawa minji kanreishū*. Vol. 5. Tokyo: Tachibana Shoin, 1986.

Mody, N. H. N. *Japanese Clocks*. Rutland, VT: C. E. Tuttle Co., 1967.

Momo Hiruyuki 桃裕行. *Rekihō no kenkyū*. Kyoto: Shibunkaku, 1990.

Morishima Chūryō 森島中良. *Kōmō zatsuwa* 紅毛雑話 [All kind of Dutch stories]. 1787. Waseda.

Morishita Tōru. "Time in Early Modern Local Community." In "The Birth of Tardiness," special issue, *Japan Review* no. 14 (2002): 65–73.

Morris-Suzuki, Tessa. "Indigenous Knowledge in the Mapping of the Northern Frontier Regions." In *Cartographic Japan*, edited by Wigen et al. Chicago: University of Chicago Press, 2016.

———. *The Technological Transformation of Japan*. Cambridge: Cambridge University Press, 1994.

Motoyasu Hiroshi 本康宏史. "*Kagahan to 19 seiki no gijutsu bunka.*" In *Endō Takanori wo chūshin ni okonowareta Kagahan no gijutsu bunka no kenkyū*, edited by Watanabe Makoto, 1–12. Toyama: Watanabe Makoto, 2006.

———. *"Kaga ni okeru karakuri zui shahon to gijutsu denpō: Ōno Daikichi no keifu wo chūshin ni." Ishikawaken Rekishi Hakubutsukan Kiyō* 11 (1998):12–42.

Mumford, Lewis. *Technics and Civilizations.* New York: Harcourt, Brace and Company, 1934.

Murai Masahiro 村井昌弘. *Ryōchi shinan* 量地指南 [Introduction to land surveying] 1733. Waseda. First published 1723.

Myers, Fred R., ed. *The Empire of Things.* Oxford: School of American Research Press, 2001.

Naganuma Kenkai 長沼賢海. *Nihon kaijishi kenkyū.* Fukuoka: Kyūshū Daigaku Shuppankai, 1976.

Nakagawa Gorōzaemon 中川五郎左衛門. *Edo kaimono hitori annai* 江戸買物獨案内 [A single guide to (all) Edo shopping]. Kobe University Digital Library.

Nakai Riken 中井履軒. *Kashoreki* 華胥暦, cited in Okada, *Meiji kaireki*, 39–40. Tokyo: Taishukan Shoten, 1994.

Nakai Takeyama 中井竹山. *Sōbōkigen* 草茅危言 [Dangerous words of grass and straw], cited in Okada, *Meiji kaireki*, 37. Tokyo: Taishukan Shoten, 1994.

Nakamura Tsukō 中村士. "The Earliest Telescope Preserved in Japan." *Journal of Astronomical History and Heritage* 11, no. 3 (2008): 203–12.

———. *Edo no Tenmongakusha Hoshizora o Kakeru*. Tokyo: Gijutsu Hyoronsha, 2008.

———. *"1820 nendai—Bakumatsuki no kikai dokei shiyou to jikoku seido,"* presented at *Higashi-ajia no koyomi to kindaika—jikangaku kokusai shimpojiumu*, November 21, 2009.

———. *"Inō Tadataka ga Setōnaikai sokuryōde shiyōshita tenmonsokki to 'yonakasokuryōnozu' no kansokuchi." Inoh Tadataka* 80 (2016): 1–10.

———. *"Kagakushi Nyūmon: bakufu tenmongata Takahashi Yoshitoki—sono shōgai, gyōseki to eikyō." Kagakushi Kenkyū* 48, no. 251 (Autumn 2009): 156–61.

———. *"Kume Eizaemon Tsūken no Nihon tenmon, sokuryōshijō niokeru ichi."* In *Kume Tsūken ni kansuru kisōteki chōsa, kenyū hōkoku: Edo no monodukuri*, edited by Matsumura, 9–10. Tokyo: Monbukagakusho, 2005.

———. *"Satō Issai no tokei kenkyū to bakumatsu tenmongata to no kōryū." Tōyō kenkyū* 171 (2009): 1–26.

———. *"Shibukawa Kagesuke cho 'jishingimontō' no seiritsu nen no suitei." Tokyo Tenmondaihō* 21, no. 2 (1988): 276–79.

Nakane Genkei 中根元圭. *Jikokuron* 時刻論 [Treatise on time]. Date unknown. Wasan.

Nakayama Shigeru 中山茂. *A History of Japanese Astronomy.* Cambridge: Harvard University Press, 1969.

———. *Nihon no Tenmongaku.* Tokyo: Iwanami Shoten, 1972.

———. *The Orientation of Science and Technology: A Japanese View.* Folkestone, UK : Global Oriental, 2009.

Nanni, Giordano. *The Colonisation of Time*. Manchester: University of Manchester Press, 2012.

Naquin, Susan. "The Forbidden City Goes Abroad: Qing History and the Foreign Exhibition of the Palace Museum, 1974–2004." *T'oung Pao* 90 (2004): 341–97.

Needham, Joseph. *Heavenly Clockwork.* Cambridge: Antiquarian Horological Society, 1960.

———. *Science and Civilization in China*. Vol. 3, *Mathematics and the Sciences of the Heavens and the Earth.* Cambridge: Cambridge University Press, 1959.

Ng, Wai-ming. *The I Ching In Tokugawa Thought and Culture.* Honolulu: Association for Asian Studies, 2000.

Nihon Gakushiin and Nihon Kagakushi Kankōkai, eds. *Meiji-zen Nihon tenmongakushi.* Tokyo: Nihon Gakujutsu Shinkōkai, 1960.

Nishikawa Joken 西川如見. *Nagasaki Yawasō* 長崎夜話草 [Night-time stories of Nagasaki]. Reprinted in *Nagasaki Sōsho*. Nagasaki: Nagasaki Shiyakusho, 1926.

———. *Ryōgi shūsetsu* 両儀集説 [Collected explanations of both models]. Date unknown. NDL.

Nishimoto Ikuko. "The Civilization of Time: Japan and the Adoption of the Western Time System." *Time and Society* 6 (1997): 237–59.

———. *Jikan ishiki no kindai: "toki wa kane nari" no shakaishi*. Tokyo: Hosei University Press, 2006.

———. "Teaching Punctuality: Inside and Outside the Primary School." In "The Birth of Tardiness," special issue, *Japan Review* no. 14 (2002): 121–34.

Nishimura Tachū 西村太沖. *Shitendai sho ki zu* 司天台諸器図 [Illustrations of various instruments in observatory]. 1805. Waseda.

Nishimura Tōsato 西村遠里, *Sokuryō sho ki zu* 測量諸器図 [Illustrations of various measuring devices]. Date unknown. NAOJ.

Nōda Chūryō 能田忠亮. *Koyomi*. Tokyo: Shibundō, 1957.

———. *Koyomi no honshitsu to sono kairyō*. Tokyo: Nihon Hōsōshuppan Kyōkai, 1943.

Norman, Donald. *The Design of Everyday Things*. New York: Doubleday, 1990.

———. *Turn Signals Are the Facial Expressions of Automobiles*. Reading, MA: Addison-Wesley, 1992.

Numata Jirō. *Western Learning*. Tokyo: Japan-Netherlands Institute, 1992.

Ōba Osamu. *Books and Boats*. Translated by Joshua A. Fogel. Portland, ME: MerwinAsia, 2012.

Oda Sachiko 小田幸子. *Wadokei zuroku*. Tokyo: Seiko Tokei Shiryokan Zō, 1994.

Oda Sachiko 小田幸子 and Sasaki, Katsuhiro 佐々木勝浩. *Zuroku wadokei*. Tokyo: Tokyo Science Museum, 1981.

Ogawa Tomotada. *Seiyō jishingi teikoku kassoku* 西洋時辰儀定刻活測 [On time measurement with Western watches]. Introduction dated 1838, main text dated 1857. Kanō.

Ogle, Vanessa. *The Global Transformation of Time*. Cambridge, MA: Harvard University Press, 2015.

Okada Masahiko. "Vision and Reality: Buddhist Cosmography in Nineteenth Century Japan." PhD diss., Stanford University, 1997.

Okada Yoshirō 岡田芳郎. *Koyomi o shiru jiten*. Tokyo: Tokyodō, 2006.

———. *Kyūreki no yomihon*. Osaka: Sogensha, 2006.

———. *Meiji kaireki: "toki" no bunmei kaika*. Tokyo: Taishukan Shoten, 1994.

———. *Nanbu Kaireki*. Tokyo: Hōsei University press, 1980.

———. *Nihon no koyomi*. Tokyo: Shin Jinbutsu Ōraisha, 1996.

Onabe Tomoko 尾鍋智子. *Zettai Tōmei no tankyū: Endō Takanori chō "Shashinjutsu" no kenkyū*. Kyoto: Shibunkaku, 2006.

Ooms, Herman. *Tokugawa Ideology*. Princeton: Princeton University Press, 1985.

Ōsaki Shōji 大崎正次. *Kinsei Nihon tenmon shiryō*. Tokyo: Hara Shobō, 1994.

Ōshima Shiran 大島芝蘭. *Tokei kō* 時計考 [Contemplation about clocks]. 1726. Gakushiin.

Ōtani Ryōkichi大谷亮吉. *Inō Tadataka*, Tokyo: Iwanami Shoten, 1917.

———. *Tadataka Inō: The Japanese Land-Surveyor*. Translated by Sugimura Kazue. Tokyo: Iwanami Shoten, 1932.

Pagani, Catherine. *Eastern Magnificence and European Ingenuity*. Ann Arbor: University of Michigan Press, 2001.

Perdue, Peter. *China Marches West*. Cambridge: Belknap Press of Harvard University Press, 2005.

Pickering, Andrew. *The Mangle of Practice*. Chicago: University of Chicago Press, 1995.
Polanyi, Michael. *Personal Knowledge*. Chicago: University of Chicago Press, 2015. First published 1958.
Postone, Moishe. *Time, Labor, and Social Domination*. New York: Cambridge University Press, 1993.
Price, Derek de Sola. "Automata and the Origins of Mechanism and the Mechanistic Philosophy." *Technology and Culture* 5 (1964): 9–23.
Riskin, Jessica. *Restless Clock*. Chicago: University of Chicago Press, 2016.
Robertson, J. Drummond. *The Evolution of Clockwork*. London: Cassell, 1931.
Rocke, Alan. *Image and Reality*. Chicago: University of Chicago Press, 2010.
Rooney, David, and James Nye. "'Greenwich Observatory Time for the Public Benefit': Standard Time and Victorian Networks of Regulation." *British Journal for the History of Science* 42, no. 1 (2009): 5–30.
Rorty, Richard. *Contingency, Irony, and Solidarity*. Cambridge: Cambridge University Press, 1989.
Saitō Kakuki 斎藤鶴磯. *Shikankō* 支干考 [Contemplations about the Shi and Kan]. 1793. NDL.
Saitō Kuniji 斉藤国治. *Kodai no jikoku seido*. Tokyo: Yuzankaku: 1995.
———. *Kotenmongaku no michi*. Tokyo: Hara Shobō, 1990.
Sakamoto Tarō et al., ed. *Nihonshoki*. Tokyo: Iwanami Bunko, 1995.
Satō Issai 佐藤一斎. *Satō Issai zenshū*. Edited by Okada Takehiko. Tokyo: Meitoku Shuppansha, 1990.
———. *Gaishi jiki zakki* 磕子時器雑記 [Records of a ticking timepiece]. 1818. Kanō.
———. *Yōsei sokujiki* 洋製測時器 [On Western-made time measurement devices].1838. Kanō.
Satō Masatsugu 佐藤政次. *Rekigakushi taizen*. Tokyo: Surugadai shuppansha, 1977.
Satō Masayuki. "Comparative Ideas of Chronology." *History and Theory* no. 3 (1991): 275–301.
Sasaki Hisao 佐々木寿男. "*Wadokei no bunken, washo (I).*" *Wadokei* no. 3 (May 1996): 1–37.
Sauter, Michael. "Clockwatchers and Stargazers." *American Historical Review* 112, no. 3 (June 2007): 685–709.
Sawada Taira 澤田平. "*Edo jidai no jishin yochigi—sono hakken to fukugen.*" *Wadokei* no. 7 (December 2000): 20–22.
———. "*Kaga no kuni no suiyōkyūgi.*" *Wadokei* no. 5 (September 2000): 5–9.
———. "*Shumisengi monogatari.*" *Wadokei* no. 41 (January 2007): 17–23.
———. "*Tentai kansokuyō wadokei [jujikan] no hatsumei to fukugen.*" *Wadokei* no. 40 (October 2005): 12–20.
Schäfer, Dagmar. *The Crafting of the 10,000 Things*. Chicago: University of Chicago Press, 2011.
Schaffer, Simon. "Astronomers Mark Time." *Science in Context* 2, no. 1 (1988): 115–45.
Screech, Timon. "Clock Metaphors in Edo Period." *Japan Quarterly* (October–December 1996): 66–75.
———. *The Lens within the Heart*. Honolulu: University of Hawai Press, 2002.
Senga Kōhei 千賀耕平. *Edo no gotokeishi: wasurerareta Nihon no kikai kōgei*. Kyoto: Ofusha, 2000.
Shiba Kōkan 司馬江漢. *Oranda Tensetsu* 和蘭天説 [Dutch explanation of heavens]. 1795. Waseda.

Shibukawa Kagesuke 渋川景佑. *Jishingi montō* 時辰義問答 [Questions and answers about chronometers]. Undated. Kanō.

———. *Kansei Rekisho* 寛政暦書 [The Book of Kansei Calendar]. 1859. NDL.

———. *Kōkai yōhō setsuyaku* 航海要法節訳 [Explanation of navigation method]. 1861. Wasan.

———. *Reigen kōbo* 霊憲候簿 [Account of divine constitution]. 1797–1846. NAOJ.

———. *Shōchū jishingi jimō* 掌中時辰儀示蒙 [Pointing out the ignorance about pocket watches]. 1823. Tokyo University Archive.

———. *Zoku kaichū funamichi kō* 続海中舟道考 [Addendum to "Study of sailing in the open sea"]. Preface from 1856. NAOJ.

Shibukawa Shunkai 渋川春海. *Tenmon keitō* 天文瓊統 [Astronomical gem string]. 1698. In *NST*, vol. 63. Tokyo: Iwanami Shoten, 1971.

Shimada Dōkan島田道桓. *Chōkenbengi* 町見辨疑 [Questions about surveying cities]. 1725. NDL

Shimada Shingo. "Social Time and Modernity in Japan." *Time and Society* 4, no. 2 (1995): 225–60.

Shinminato Museum, ed. *Etchū no ijin Ishiguro Nobuyoshi*. Shinminato: Shinminato Museum, 2001.

Shirahata Yōzaburō. "The Development of Japanese Botanical Interest and Dodonaeus' Role: From Pharmacopoeia to Botany and Horticulture." In *Dodonaeus in Japan*, edited by Willy vande Walle and Kazuhiko Kasaya, 263–79. Leuven: Leuven University Press, 2001.

Shirane Haruo. *Japan and the Culture of the Four Seasons*. New York: Columbia University Press, 2012.

Shizuki Tadao 志筑忠雄. *Sakokuron* 鎖国論[Treatise on the locked country]. 1801. NDL.

Shōtei Shujin 松亭主人. *Rekijitsu chūkai* 暦日註解 [Explanation of calendrical pointers]. 1848–1854. NDL.

Shūyūdō Shujin 尚友堂主人. *Waei tsūin iroha benran* 和英通韻以呂波便覧 [English-Japanese basic guide]. 1868. NDL.

Siebold Philipp Franz von. *Nippon, Archiv zur Beschreibung von Japan*. Wurzburg: Woerl, 1897.

Sivin, Nathan. "Cosmos and Computation in Early Chinese Mathematical Astronomy." *T'oung Pao* 55 (1969): 1–73.

———. *Granting the Seasons*. New York: Gardners Books, 2008.

Smith, Thomas. *Native Sources of Japanese Industrialization, 1750–1920*. Berkeley: University of California Press, 1988.

Sōma Mitsuru et al. "Units of Time in Ancient China and Japan." *Publications of Astronomical Society of Japan* 56 (2004): 887–904.

Sugimoto Masayoshi, and David Swain. *Science and Culture in Traditional Japan*. Rutland, VT: Charles E. Tuttle Co., 1989.

Sugimoto Tsutomu 杉本つとむ. *Edo jidai Rangogaku no seiritsu to sono tenkai*. Tokyo: Waseda Daigaku Shuppanbu, 1976–1982.

———. *Edo no honyakutachi*. Tokyo: Waseda Daigaku Shuppanbu, 1995.

———. *Tenmon rekigaku shoshū*. Tôkyô: Waseda Daigaku Shuppanbu, 1996–1997.

Sugita Genpaku. *Dawn of Western Science in Japan. Rangaku Kotohajime*. Translated by Tomio Ogata. Tokyo: Hokuseido Press, 1969.

Sumita Shōichi 住田正一. *Kaiji shiryō sōsho*. Tokyo : Ganshōdō Shoten, 1929–1931.

Suzuki Junko. "Seeking Accuracy: The First Modern Survey of Japan's Coast." In *Carto-*

graphic Japan, edited by Wigen et al., 129–32. Chicago: University of Chicago Press, 2016.

Suzuki Kazuyoshi et al. "*Kume Tsūken ni miru edo jidai no kagakugijutsu.*" In *Kume Tsūken ni kansuru kisōteki chōsa, kenyū hōkoku: Edo no monodukuri*, edited by Matsumura, 3–6. Tokyo: Monbukagakusho, 2005.

———. "*Mannnendokei no kikō kaimei sono 1: wadokei.*" *Proceedings of the Annual Meeting of the Japan Society of Mechanical Engineers* 5 (2005):51-52.

———. "Mannnendokei no kikō kaimei sono 2: tenkyūgi." *Proceedings of the Annual Meeting of the Japan Society of Mechanical Engineers* 5 (2005):53-54.

———. "Mechanism of '*Man-non dokei*,' a Historic Perpetual Chronometer. Part 2: Power Supply," 1–6. 12th IFToMM World Congress, Besançon, France, June 18–21, 2007.

Suzuki Nei 鈴木寧. *Edo Meiji shosetsu no tokei ga*. Hirosaki: Roku no fue mamehon no kai, 1999.

Suzaki Tetsuji 洲崎哲二. *Nishimura Tachū jiseki*. Jōhanamachi: Nishimura Tachū sensei hikensetsu kyousankai, 1934.

Tagaya Kanchusen 多賀谷環中仙. *Karakuri Kinmō Kagamigusa* 璣訓蒙鑑草 [Encyclopedia of automata]. 1730. Waseda.

Tachibana Nankei 橘南谿. *Tōyūki* 東遊記 [Record of travels to the East]. Undated. Waseda.

Tadano Makuzu, Janet R. Goodwin, Bettina Gramlich-Oka, Elizabeth A. Leicester, Yuki Terazawa, and Anne Walthall. "Solitary Thoughts—A Translation of Tadano Makuzu's *Hitori Kangae* (2)." *Monnumenta Nipponica* 56, no. 2 (2001): 173–95.

Takahashi Hyōei 高林兵衛. *Tokei hattatsu shi*. Tokyo: Tōyō shuppansha, 1924.

Takahashi Kageyasu 高橋景保. *Rōshia shinto peteruburugu no zu* 魯西亜新都ペテルブルグ之図 [A map of the new Russian capital of St. Petersburg]. Hazama.

———. *Shinsen sōkai zenzu* 新鐫総界全図 [New map of the entire world]. Waseda.

———. *Shintei Bankoku zenzu* 新訂万国全図 [New map of the myriad countries]. Waseda.

Takahashi Satoshi 高橋敏. *Edo no soshō*. Tokyo: Iwanami Shoten, 1996.

Takahashi Yoshitoki 高橋至時. *Angeria reki kō* 諳厄利亜暦考 [Study of the English almanac]. 1801. Wasan.

———. *Kaichū funamichikō* 海中舟道考 [Thoughts on seafearing routes].1803. Gakushiin.

———. *Seiyōjin Rarande rekisho kanken* 西洋人ラランデ暦書管見 [A brief glance at Lalande's book of astronomy]. In *NST*, vol. 65. Tokyo: Iwanami Shoten, 1972.

Takata Shigehiro 高田茂廣. *Kinsei Chikuzen kaijishi no kenkyū*. Tokyo: Bunken Shuppan, 1993.

Takeda Izumo et al. *Kanadehon chūshingura* 仮名手本忠臣蔵 [The tale of the forty-seven *rōnin*]. 1747. Tokyo: Iwanami Shoten, 1937.

Takeuchi Shinichiro 竹内慎一郎. *Chizu no Kioku: Ino Tadataka Etchū sokuryō ki*. Toyama: Katsura Shobo, 1999.

Tamura Takeo 田村竹男. *Ibaraki no tokei*. Tsukuba: Furusato Bunko, 1990.

———. "*Ino Tadataka no suiyōkyūgi ni tsuite.*" *Wadokei* no. 1 (June 1992): 32–41.

Tanaka Hisashige 田中久重 (signed Fukube Ōmi 服部近江). *Shoki kōan zu* 諸器考案図 [Sketches of designes of various devices]. Undated. Tokyo Edo Museum Archive.

Tanaka, Stefan. *New Times in Modern Japan*. Princeton, NJ: Princeton University Press, 2004.

Tatsukawa Shōji 立川昭二 et al., ed. *Zusetsu karakuri: asobi no hyakka*. Tokyo: Kawade Shobo Shinsha, 2002.

Terajima Ryōan 寺島良安. *Wakan Sansai Zue* 和漢三才図絵 [Illustrated Sino-Japanese encyclopedia]. 1713. Kyushu University Digital Archive.

Terrall, Mary. *The Man Who Flattened the Earth*. Chicago: University of Chicago Press, 2002.

Thatcher, Deane E. "Instruments and Observation at the Imperial Astronomical Bureau during the Ming Dynasty." *Osiris* 9 (1994): 127–40.

Thompson, E. P. "Time, Work-Discipline and Industrial Capitalism." *Past and Present* 38 (December 1967): 56–97.

Titsingh, Isaac. *Illustrations of Japan*. London: Printed for R. Ackermann, 1822. Princeton Rare Books Collection.

Tokugawa Jikki 徳川実記. [Records of Tokugawa House]. In *Kokushi taikei*, vol. 48. Tokyo: Yoshikawa Kōbunkan, 1934.

Tsukamoto Aketake 塚本明毅. *Tsukamoto Aketake Kengi* 塚本明毅建議 [Tsukamoto Aketake proposal], cited in Okada, *Meiji kaireki*, 126–28. Tokyo: Taishukan Shoten, 1994.

Truit, Elly. "The Incarnation of Time." In *L'automate: modèle, métaphore, machine, merveille*, edited by Galliard et al., 365–78. Bordeaux: Presses univérsitaire de Bordeaux, 2012.

Tucker, Mary Evelyn. "Religious Dimensions of Confucianism: Cosmology and Cultivation." *Philosophy East and West* 48, no. 1 (January 1998): 5–45.

Tsunoyama Sakae 角山栄. *Tokei no shakaishi*. Tokyo: Chūkō Shinsho, 1984.

Tsutsumi Ichirō 堤一郎. "*Tokei wottchingu.*" *Nihon Tokei Gakkaishi* 144 (1993): 76–80.

Toda Teruyoshi 戸田光良. "*Edo jidai no tenmon tokei: suiyōkyūgi monogatari.*" *Wadokei*, no. 6 (September 2000): 31–54.

Tokyo Kagaku Hakubutsukan 東京科学博物館. *Edo Jidai no Kagaku*. Tokyo: Meicho Kankoukai, 1976.

Turner, Anthony J. *Of Time and Measurement*. Brookfield, VT: Variorum, 1993.

Turner, Stephen. *Understanding the Tacit*. New York: Routledge, 2014.

Uehara Hisashi 上原久. *Takahashi Kageyasu no kenkyū*. Tokyo: Kōdansha, 1977.

Uehara Hisashi et al., eds. *Tenmon rekigaku shoka shokanshū*. Tokyo: Kodansha, 1981.

Umeda Chihiro 梅田千尋. *Kinsei onmyōdō soshiki no kenkyū*. Tokyo: Yoshikawa Kōbunkan, 2009.

Unno Kazutaka 海野一隆. *Nihonjin no daichizō: Seiyō chikyūsetsu no juyō o megutte*. Tokyo: Taishūkan Shoten, 2006.

Urai Sachiko 浦井幸子. *Edo no jikoku to toki no kane*. Tokyo: Iwata shoin, 2002.

Vande Walle, W. F. "Linguistics and Translation in Pre-Modern Japan and China." In *Dodonaeus in Japan*, edited by Willy vande Walle and Kazuhiko Kasaya, 12–13. Leuven: Leuven University Press, 2001.

Verbiest, Ferdinand, and Liu Yunde 劉 蘊德. *Lingtai yixiang zhi* 靈台儀象志 [History of instruments used in the (imperial) observatory]. Originally written in 1674. Tokugawa period reprint. Waseda.

Voskhul, Adelheid. *Androids in the Enlightenment*. Chicago: University of Chicago Press, 2013.

Vygotskii, Lev. *Myshlenie i rech*. Moskva: Labirint, 1996. Originally published in 1934.

Walker, Brett. *The Conquest of Ainu Lands*. Berkeley: University of California Press, 2001.

———. "Mamiya Rinzō and the Cartography of Empire." In *Cartographic Japan*, edited by Wigen et al. Chicago: University of Chicago Press, 2016.

———. "Mamiya Rinzō and the Japanese Exploration of Sakhalin Island: Cartography and Empire." *Journal of Historical Geography* 33, no. 2 (April 2007): 283–313.

Watanabe Ichirō 渡辺一郎. *Inō Tadataka no chizu wo yomu*. Tokyo: Kawade shobo, 2000.

Watanabe Makoto 渡辺誠, ed. *Endō Takanori wo chūshin ni okonawareta Kagahan no gijutsubunka no kenkyū*. Tokyo: Seika Hōkoku, 2006.

Watanabe Shin 渡辺慎. *Ryōchiden shuroku* 量地伝皆録 [All methods of land surveying]. 1831. NDL.

Watanabe Toshio 渡辺敏夫. *"Edojidaino rekihon hankōsū." Wadokei Kenkyū* no. 1 (1992): 15–24.

———. *Kinsei nihon kagakushi to Asada Gōryū*. Tokyo: Yuzankaku Publishing, 1983.

———. *Kinsei Nihon Tenmongakushi: Kansatsu gijutsu*. Tokyo: Koseisha Koseikaku, 1987.

———. *Kinsei Nihon Tenmongakushi: Tsūshi*. Tokyo: Koseisha Koseikaku, 1986.

———. *Koyomi to Tenmon*. Tokyo: Kôseisha Kôseikaku, 1966.

———. *Nihon no koyomi*. Tokyo: Yūzankaku, 1993.

———. *Tenmon rekigakushijō ni okeru Hazama Shigetomi to sono ikka*. Kyoto: Yamaguchi Shoten, 1943.

Waugh, Albert E. *Sundials: Their Theory and Construction*. New York: Dover, 1973.

Whitrow, G. J. *The Nature of Time*. New York: Holt, Rinehart and Winston, 1973.

Wigen, Kären. *A Malleable Map*. Berkeley: University of California Press, 2010.

Wilson, George Macklin. "Time and History in Japan." *American Historical Review* 85, no. 3 (June 1980): 557–71.

Winichakul, Thongchai. *Siam Mapped*. Honolulu: University of Hawaii Press, 1994.

Wise, Norton M., ed. "Mediating Machines." *Science in Context* 2, no. 1 (1988): 77–113.

———. *The Values of Precision*. Princeton: Princeton University Press, 1995.

Wood, Denis. *The Power of Maps*. New York: Guilford Press, 1992.

Yamagata Bantō 山片蟠桃. *Yume no shiro* 夢之代 [Instead of Dreams]. 1802. NDL.

Yamaguchi Ryuji 山口隆二. *Nihon no tokei*. Tokyo: Nihon Hyōronsha, 1942.

Yamamoto Shinsaburō 山本沈三郎. *Shichijūnikō idōkō* 七十二候異同攷 [Examination of similarities and differences in the seventy two seasons]. 1857. Waseda.

Yanagiwa Shunsan 柳河春三. *Seiyō tokei benran*西洋時計便覧 [Basic guide for Western clocks]. Kanō.

Yonemoto Marcia. *Mapping Early Modern Japan*. Berkeley: University of California Press, 2003.

Yoshida Tadashi. "The *Rangaku* of Shizuki Tadao." PhD diss., Princeton, 1974.

———. *"Takahashi Yoshitoki to seiyou tenmongaku." Astronomical Herald* no. 5 (2005): 291–99.

Zhang, Qiong. "Demystifying Qi: The Politics of Cultural Translation and Interpretation in the Early Jesuit Mission to China." In *Tokens of Exchange*, edited by Lidia Liu, 74–106. Durham: Duke University Press, 1999.

———. *Making the New World Their Own*. Leiden: Brill, 2015.

Zöllner, Reinhard. *Japanische Zeitrechnung: ein Hadbuch*. München: Iudicium, 2003.

INDEX

Page numbers in italics refer to figures.

Studies of the Weatherhead East Asian Institute
Columbia University

SELECTED TITLES

(Complete list at: http://weai.columbia.edu/publications/studies-weai/)

Resurrecting Nagasaki: Reconstruction and the Formation of Atomic Narratives, by Chad Diehl. Cornell University Press, 2018.

Promiscuous Media: Film and Visual Culture in Imperial Japan, 1926–1945, by Hikari Hori. Cornell University Press, 2018.

The End of Japanese Cinema: Industrial Genres, National Times, and Media Ecologies, by Alexander Zahlten. Duke University Press, 2017.

The Chinese Typewriter: A History, by Thomas S. Mullaney. The MIT Press, 2017.

Mobilizing Without the Masses: Control and Contention in China, by Diana Fu. Cambridge University Press, 2017.

Forgotten Disease: Illnesses Transformed in Chinese Medicine, by Hilary A. Smith. Stanford University Press, 2017.

Food of Sinful Demons: Meat, Vegetarianism, and the Limits of Buddhism in Tibet, by Geoffrey Barstow. Columbia University Press, 2017.

Aesthetic Life: Beauty and Art in Modern Japan, by Miya Mizuta Lippit. Harvard University Asia Center, 2017.

Youth For Nation: Culture and Protest in Cold War South Korea, by Charles R. Kim. University of Hawaii Press, 2017.

Socialist Cosmopolitanism: The Chinese Literary Universe, 1945–1965, by Nicolai Volland. Columbia University Press, 2017.

Yokohama and the Silk Trade: How Eastern Japan Became the Primary Economic Region of Japan, 1843–1893, by Yasuhiro Makimura. Lexington Books, 2017.

The Social Life of Inkstones: Artisans and Scholars in Early Qing China, by Dorothy Ko. University of Washington Press, 2017.

Darwin, Dharma, and the Divine: Evolutionary Theory and Religion in Modern Japan, by G. Clinton Godart. University of Hawaii Press, 2017.

Dictators and Their Secret Police: Coercive Institutions and State Violence, by Sheena Chestnut Greitens. Cambridge University Press, 2016.

The Cultural Revolution on Trial: Mao and the Gang of Four, by Alexander C. Cook. Cambridge University Press, 2016.

Inheritance of Loss: China, Japan, and the Political Economy of Redemption after Empire, by Yukiko Koga. University of Chicago Press, 2016.

Homecomings: The Belated Return of Japan's Lost Soldiers, by Yoshikuni Igarashi. Columbia University Press, 2016.

Samurai to Soldier: Remaking Military Service in Nineteenth-Century Japan, by D. Colin Jaundrill. Cornell University Press, 2016.

The Red Guard Generation and Political Activism in China, by Guobin Yang. Columbia University Press, 2016.

Accidental Activists: Victim Movements and Government Accountability in Japan and South Korea, by Celeste L. Arrington. Cornell University Press, 2016.

Ming China and Vietnam: Negotiating Borders in Early Modern Asia, by Kathlene Baldanza. Cambridge University Press, 2016.

Ethnic Conflict and Protest in Tibet and Xinjiang: Unrest in China's West, coedited by Ben Hillman and Gray Tuttle. Columbia University Press, 2016.

One Hundred Million Philosophers: Science of Thought and the Culture of Democracy in Postwar Japan, by Adam Bronson. University of Hawaii Press, 2016.

Conflict and Commerce in Maritime East Asia: The Zheng Family and the Shaping of the Modern World, c. 1620–1720, by Xing Hang. Cambridge University Press, 2016.

Chinese Law in Imperial Eyes: Sovereignty, Justice, and Transcultural Politics, by Li Chen. Columbia University Press, 2016.

Imperial Genus: The Formation and Limits of the Human in Modern Korea and Japan, by Travis Workman. University of California Press, 2015.

Yasukuni Shrine: History, Memory, and Japan's Unending Postwar, by Akiko Takenaka. University of Hawaii Press, 2015.

The Age of Irreverence: A New History of Laughter in China, by Christopher Rea. University of California Press, 2015.

The Knowledge of Nature and the Nature of Knowledge in Early Modern Japan, by Federico Marcon. University of Chicago Press, 2015.

The Fascist Effect: Japan and Italy, 1915–1952, by Reto Hofmann. Cornell University Press, 2015.

The International Minimum: Creativity and Contradiction in Japan's Global Engagement, 1933–1964, by Jessamyn R. Abel. University of Hawaii Press, 2015.

Empires of Coal: Fueling China's Entry into the Modern World Order, 1860–1920, by Shellen Xiao Wu. Stanford University Press, 2015.